Gender Differences in Susceptibility to Environmental Factors: A Priority Assessment

Workshop Report

Valerie Petit Setlow, C. Elaine Lawson, and
Nancy Fugate Woods, *Editors*

Committee on Gender Differences in Susceptibility to
Environmental Factors

Division of Health Sciences Policy

INSTITUTE OF MEDICINE

NATIONAL ACADEMY PRESS
Washington, D.C. 1998

NATIONAL ACADEMY PRESS • 2101 Constitution Avenue, N.W. • Washington, DC 20418

NOTICE: The project that is the subject of this report was approved by the Governing Board of the National Research Council, whose members are drawn from the councils of the National Academy of Sciences, the National Academy of Engineering, and the Institute of Medicine. The members of the committee responsible for the report were chosen for their special competences and with regard for appropriate balance.

The Institute of Medicine was chartered in 1970 by the National Academy of Sciences to enlist distinguished members of the appropriate professions in the examination of policy matters pertaining to the health of the public. In this, the Institute acts under both the Academy's 1863 congressional charter responsibility to be an adviser to the federal government and its own initiative in identifying issues of medical care, research, and education. Dr. Kenneth I. Shine is president of the Institute of Medicine.

Support for this project was provided by the Department of Health and Human Services, National Institutes of Health (NIH) Office of Research on Women's Health (Award No. 1-OD-4-2139), with contributions from the National Institute of Environmental Health Sciences, NIH, the National Institute of Child Health and Human Development, the Office of Women's Health, the Centers for Disease Control and Prevention, and the Office of Research and Development, Environmental Protection Agency. The views presented in this report are those of the Committee on Gender Differences and are not necessarily those of the funding organizations.

International Standard Book No. 0-309-06423-6

Additional copies of this report are available for sale from the National Academy Press, 2101 Constitution Avenue, N.W., Box 285, Washington, DC 20055; call (800) 624-6242 or (202) 334-3313 (in the Washington metropolitan area), or visit the NAP's on-line bookstore at **http://www.nap.edu**.

For more information about the Institute of Medicine, visit the IOM home page at **http://www2.nas.edu/iom**.

Copyright 1998 by the National Academy of Sciences. All rights reserved.

Printed in the United States of America
First Printing, March 1998
Second Printing, November 1998

COMMITTEE ON GENDER DIFFERENCES IN SUSCEPTIBILITY TO ENVIRONMENTAL FACTORS

NANCY FUGATE WOODS (*Chair*),* Director, Center for Women's Health Research and Professor, Family and Child Nursing, School of Nursing, University of Washington
EULA BINGHAM,* Professor, Environmental Health, University of Cincinnati
KIM BOEKELHEIDE, Professor, Department of Pathology, Brown University
DENISE FAUSTMAN, Director of Immunobiology Laboratories, Massachusetts General Hospital, and Harvard Medical School
STEPHEN H. SAFE, Distinguished Professor, Department of Veterinary Medicine, Physiology, and Pharmacology, Texas A&M University
DAVID H. WEGMAN, Professor and Chair, Department of Work Environment, University of Massachusetts, Lowell

IOM Health Sciences Policy Board Member/Committee Liaison

ADA SUE HINSHAW,* Dean, School of Nursing, University of Michigan

Institute of Medicine Staff

VALERIE PETIT SETLOW, Director, Division of Health Sciences Policy and Project Director
C. ELAINE LAWSON, Research Associate and Project Codirector
LINDA A. DEPUGH, Administrative Assistant and Project Assistant
JAMAINE TINKER, Financial Associate

---

*Member, Institute of Medicine.

Preface and Acknowledgments

Women's and men's health differ in a variety of ways. Women have greater longevity than men, but during their lifespan women experience more morbidity. Scientists have investigated explanations for these differences, pursuing explorations of biological differences, such as those linked to the X chromosome and those modulated by sex steroids (e.g., immune response). Other scholars have studied the differential socialization of girls and boys with respect to risk-taking behavior, sophistication about health and health-seeking behavior, and the social roles women and men play in their occupations and in their homes. Still others have examined sources of stress in women's and men's lives that might account for differences in health and disease patterns. The most likely explanations accounting for women's and men's different health experiences are complex and multivariate and may include differences in each gender's unique susceptibility to factors in their environments.

Recognizing the complexity of the topic, the Committee on Gender Differences in Susceptibility to Environmental Factors undertook the study within a framework that incorporated distinctions between sex and gender and defined environment in its broadest sense—inclusive of physical, biological, social, and cultural dimensions. As an initial step toward these fundamental understandings, our committee was assembled to review existing information, discuss issues with a larger group of interested individuals, and make recommendations for an initial set of priorities for work in this area.

Although the committee bears responsibility for the conclusions and recommended priorities in this report, I would be seriously remiss if I failed to acknowledge the contributions of many others to both the planning and conduct of the committee's activities. First, I owe thanks to the sponsors of this activity for posing the questions and initiating this study. Special thanks go to Dr. Vivian Pinn, director of the Office for Research on Women's Health (ORWH) at the

National Institutes of Health (NIH), for initiating this effort and to then deputy director, Dr. Anne Bavier, who served ably as the task leader in the initial stages of this project. I would also like to acknowledge the skillful work of Joyce Rudick, acting deputy director, ORWH, who graciously stepped into the role of task leader after Dr. Bavier's departure from ORWH.

One of the features of this small but important project was its multiagency sponsorship and interest. Joining ORWH in support of this project were Dr. Anne Sassaman, director, Division of Extramural Research and Training, and Dr. Gwen W. Collman, scientific programs administrator, both of the National Institute of Environmental Health Sciences, NIH, and Dr. Yvonne Maddox, deputy director, National Institute of Child Health and Human Development, NIH. The committee thanks Dr. Lynn Goldman, assistant administrator for prevention, pesticide, and toxic substances, and Dr. Margaret Chu, toxicologist from the Environmental Protection Agency. The committee also thanks Dr. Wanda Jones, associate director for women's health at the Centers for Disease Control and Prevention. Each of these agency representatives provided us with materials and helpful ideas during the course of this activity. In addition, the committee extends its appreciation and thanks to the many other federal agency representatives who were contacted for information throughout the activity and who attended the workshop.

The workshop speakers shared their extensive expertise and provided the committee with thoughtful insights and ideas during the discussion period. They also helped the committee shape its priority recommendations. Therefore, I thank our excellent speakers: Greg Cosma, assistant professor, Department of Environmental Health, Colorado State University; S. Katharine Hammond, associate professor of environmental health sciences, School of Public Health, University of California at Berkeley; Kenneth S. Korach, scientific program director, Environmental Diseases and Medicine Program, National Institute of Environmental Health Sciences, NIH; Shiriki Kumanyika, professor and chair in the Department of Human Nutrition and Dietetics, University of Illinois at Chicago; Bill Lasley, professor of reproductive medicine, Institute for Toxicology and Environmental Health, University of California at Davis; Peter N. Riskind, chief of neuroimmunology, Massachusetts General Hospital, Boston; and Jeanne Stellman, deputy chair, Department of Health Policy and Management, Columbia University School of Public Health, New York City.

Others who contributed to the work of the committee are Paul Phelps, consultant and writer who listened with us throughout the workshop and prepared a draft report of the workshop that captured the essence of the discussions. I also thank IOM senior project officer, Carolyn Fulco, for her critical and thoughtful review of earlier manuscripts; Claudia Carl, for her careful work in shepherding the report through review; Michael Edington, for his assistance in the publication of the document; and Ted Cron, our copy editor.

As the committee chair, I am acutely aware of the contributions that the committee staff has made to the success of the study. Special thanks go to Linda DePugh, administrative assistant, who made travel arrangements and meetings as comfortable as possible and provided outstanding administrative support at the meetings and in the production of the report; to Jamaine Tinker, financial associate, for her masterful management of limited resources; to Elaine Lawson, research associate, who was instrumental in the early work of the committee in organizing materials, developing the initial analysis of the sponsoring agencies' research portfolio, and helping the committee identify speakers for the workshop; and to Valerie Setlow, division director, who provided her adept professional support to the committee throughout its tasks, report finalization, and review.

I would like to acknowledge the individual and collective efforts of the committee members. It seemed, *a priori*, that not all the questions, let alone all the answers, could emerge from such a small group. Yet, each member of the group assumed his or her tasks seriously and helped develop a very thoughtful agenda for agency action. It was a pleasure to have worked with this group of busy but unselfish professionals, who volunteered their time to share their knowledge and advice with the larger scientific community. In sum, their advice provides a good first step toward a fuller understanding of the unique and differential susceptibilities of women to environmental factors.

This report has been reviewed by individuals chosen for their diverse perspectives and technical expertise, in accordance with procedures approved by the National Research Council's Report Review Committee. The purpose of this independent review is to provide candid and critical comments that will assist the authors and the Institute of Medicine in making the published report as sound as possible and to ensure that the report meets institutional standards for objectivity, evidence, and responsiveness to the study charge. The content of the review comments and draft manuscript remain confidential to protect the integrity of the deliberative process.

On behalf of the Institute of Medicine, I wish to thank the following individuals for their participation in the review of this report: Mary Ellen Avery, M.D., professor of pediatrics, Harvard Medical School; Brigid Hogan, Ph.D., investigator, Howard Hughes Medical Institute, Department of Cell Biology, Vanderbilt University School of Medicine; Maria New, professor and chair, Department of Pediatrics, and chief, Pediatric Endocrinology, New York Hospital, New York City; Michael Paolisso, Ph.D., systems professor of anthropology, Department of Anthropology, University of Maryland at College Park; Ellen K. Silbergeld, Ph.D., director, Program in Human Health and the Environment, University of Maryland at Baltimore; Helen Rodriguez-Trias, M.D., codirector, Pacific Institute for Women's Health, Western Consortium for Public Health, Los Angeles. While these individuals have provided many constructive comments and suggestions, responsibility for the final content of this report rests solely with the Committee on Gender Differences in Susceptibility to Environmental Factors and the Institute of Medicine.

A final comment: The principal focus of this report is on women. However, it lays the groundwork for the most logical next step: an evaluation of the analytical steps required to move to a gender-focus analysis, one that would involve health outcomes for women *and* men. Work in the international field of women in development suggests that the most powerful analysis is to consider both women and men, and the relations and interactions between them in terms of particular outcomes, such as health. Women's health-seeking behavior and their ability to respond to environmental risk will in part be conditioned by their relationships with men, both from a cultural role perspective (i.e., what is appropriate for women and men to do) and what they actually do (behavior). Building a constituency of researchers and policymakers for gender differences in susceptibility to environmental factors will be fostered by a more inclusive and comparative focus.

Nancy Fugate Woods
Chair

Contents

EXECUTIVE SUMMARY 1

WORKSHOP REPORT 3
    Organizing Meeting, 4
    Workshop, 5
    Summary of Discussions from the Workshop, 7
    Priority Assessment: Conclusions and Recommendations, 10
    Recommendations, 12
    Closing Remarks, 23

APPENDIXES
A   Analysis of Agency Research Portfolios, 25
B   Summary of Workshop Presentations, 33
C   Acronyms, Abbreviations, and Glossary, 62
D   Workshop Agenda, Speakers, and Participants, 64
E   Biographies of Workshop Speakers, 69
F   References and Suggested Reading, 72
G   Committee Biographies, 76

Tables and Figures

TABLES

1 Types of Information Received from Sponsoring Agencies and Reviewed by the Committee, 4
2 NRC Reports with Information and Recommendations Related to Environmental Health Research and Training, 6
3 Additional Factors to Include in Broader Definitions, 11
4 Summary of Recommended Priorities, 24

B-1 Recommended Dietary Allowances 40
B-2 Dietary Reference Intakes,, 41
B-3 Synergistic Effect of Multiple Agents: Cigarette Smoking and Occupational Exposure to Asbestos, 45
B-4 Influences of Gender, Development, and Ethanol Consumption on Benzene's Effect on Erythroid Colony-Forming Units, 47
B-5 Gender Differences in Cancer Susceptibility: Human Studies/ Environmental Exposures, 61

FIGURES

A-1 Percentage of research portfolio by category, 27

B-1 Industrial employment of women and men by sector, 1992, 36
B-2 Setting nutrition monitoring priorities, 39
B-3 Occupational injuries, 45
B-4 Environmental tobacco smoke in the home, 46
B-5 Environmental tobacco smoke in the workplace, 47
B-6 Smoking and the risk of cancer among postmenopausal women, 49

B-7 Geographic localization of multiple sclerosis in the world, 50
B-8 Estrogen protective effects against TCDD, 58
B-9 TCDD and tyrosine kinase activity, 59
B-10 Changes that occur as a result of exposure, 60

Gender Differences in Susceptibility to Environmental Factors:
A Priority Assessment

Executive Summary

In 1996, the Office for Research on Women's Health, National Institutes of Health (NIH), requested that the Institute of Medicine (IOM) conduct a workshop study to review the current research programs of the National Institutes of Health (NIH), the Centers for Disease Control and Prevention (CDC), and the Environmental Protection Agency (EPA) that are devoted to women's health. The purpose of this activity was to identify the state of knowledge regarding gender differences in susceptibility to environmental factors and make recommendations about promising areas of inquiry that may profit from interagency coordination.

In order to do this, the committee reviewed a variety of research reports, publications, and journal articles, as well as relevant project summaries of funded research of the NIH, CDC, and EPA. Based on this literature review and analysis of existing research, the committee conducted a workshop that focused on three questions:

1. What areas within the existing portfolio are likely to yield information appropriate to this topic? What are the gaps in knowledge that warrant future research?

2. Are there research strategies and priorities for addressing the gaps in knowledge?

3. What other strategies, including interagency coordination, might improve the prospect of developing knowledge that will identify gender differences in susceptibility to environmental factors?

The committee concluded that for the purpose of promoting interagency strategies, a yearly workshop should be held. A second conclusion was that additional factors, such as ergonomic, behavioral, and cultural need to be included in the definitions of "gender," "environment," and "susceptibility." The committee made recommendations in three general areas:

1. *Research on exposures* to include a broader definition of terms; more occupational data elements; multiple-exposure data; research across lifespans and during critical periods; development of animal models; and identification of cultural and historical factors.

2. *Basic research* to include studies on environmental contributions and biological causes for gender differences; gender differences in disease outcomes; metabolic and hormonal differences; genetic markers of susceptibility; and translational research.

3. *Research policy* to include presenting annual workshops; fostering institutional changes; increasing the number of sponsoring agencies; funding long-term prospective studies; encouraging public and private cofunding; improving access to and content of national databases; and devising strategies for research resource protection and utilization.

Workshop Report

A variety of sources, such as morbidity and mortality data and health care utilization data, point to differences in health status between men and women. Some of these distinctions are thought to be associated also with race, ethnicity, and/or socioeconomic status. While critically important, these factors are not the subject of this report. Other distinctions are thought to be solely based on sex,[1] but there is growing awareness that the environment and environmental factors may play a role in creating health status differences between men and women. Various factors, such as genetics and hormones, may account for gender differences in susceptibility to environmental factors. In the development of approaches to disease prevention and health promotion, to behavioral and medical interventions, or to the initiation of research strategies, many have come to realize that special consideration must be given to health effects that are either gender-specific to or are overrepresented in women because of environmental factors such as occupation, behavior, lifestyle, hobbies, reproductive status, or physical activity. This latter series of issues is the focus of this report.

Many activities have been initiated on issues concerning women's health, and many federal agencies now have programs to address various aspects of health outcomes in women. However, there is ample room for newer opportunities for coordination and prioritization of research to answer questions about sex differences in susceptibility to environmental factors or gender variation in disease expression. Identification and clarification of the real or perceived gaps

---

[1] For the purpose of this report, sex is generally used to designate chromosomal or biologic phenomena linked to having one or two X chromosomes, whereas gender is used when referring to the social expression of living as a man or woman.

in knowledge may assist policymakers in planning future research initiatives and in interagency coordination.

Thus, in 1996, the Office for Research on Women's Health, National Institutes of Health (NIH), requested that the Institute of Medicine (IOM) conduct a workshop study to review the current research programs of the National Institutes of Health (NIH), the Centers for Disease Control and Prevention (CDC), and the Environmental Protection Agency (EPA) that are devoted to women's health. The purpose of this activity was to identify the state of knowledge regarding gender differences in susceptibility to environmental factors and make recommendations about promising areas of inquiry that may profit from interagency coordination.

In response to this request, the IOM formed a Committee on Gender Differences in Susceptibility to Environmental Factors. The committee included experts in environmental and occupational health and medicine, health sciences policy, epidemiology and public health, risk assessment, endocrinology, immunology, toxicology, and women's health. The committee met twice during the course of the study and held a scientific workshop in May 1997.

ORGANIZING MEETING

The first meeting of the committee had three purposes:

1. **Review the charge of the committee.** The phrase "gender differences" implies assessing differences and similarities between men and women; however, discussions with relevant agency representatives indicated a need to focus on women's unique susceptibility. Therefore, the committee refined its charge to focus on this latter aspect: that is, the *identification of areas in which research and policy initiatives could address women's differential susceptibility to environmental factors*.

2. **Review the existing research related to the topic of the study.** The committee reviewed a variety of materials related to environmental health research, gender differences, and environmental susceptibility. An abbreviated list of these materials, resources, and sources of information is in Table 1.

TABLE 1 Types of Information Received from Sponsoring Agencies and Reviewed by the Committee

Review of Scientific Literature
Relevant IOM/NRC Reports (see Table 2)
Agency Mission Statements and Program Overviews
Agency Strategic Planning Materials
Agency Abstracts of Active Research Projects

Based on a review of the literature and of relevant IOM/NRC reports (see Table 2), the committee understood that substantial research work is being done in environmental health research that would have an impact on understanding gender differences and environmental susceptibilities. Key areas of research include work on environmental estrogens, multiple chemical exposures, gender differences in response to toxic substances, allergens, and autoimmune and other immune responses to environmental factors. However, most of this work does not focus on gender differences.

Against this broader context of research, the committee reviewed the missions, programs, and strategic plans of the sponsoring agencies. Finally, the committee reviewed abstracts of the sponsoring agencies' research portfolios. The goal of this review was to understand what research was currently being conducted and identify areas in which future research would be useful (Appendix A).

3. **Develop an agenda for the workshop.** The committee began to outline the areas from which to gather more information. To do this, the committee and staff identified other experts who would provide presentations on various related topics and promising areas of research. Relevant data sought by the committee revealed gaps in knowledge or offered new information that displayed specific gender differences in disease initiation, progression, or outcome.

WORKSHOP

Because of the extensive information reviewed at the organizational meeting, the committee designed the workshop to review broad aspects of environmental exposure that would be common among women. These exposures were thought to occur in a variety of ways: in different settings (e.g., the home, the workplace); through different routes (e.g., foods), because of different activities (e.g., societal roles, chores, hobbies), or because of unique or critical times in the lifespan. A second design approach for the workshop was to review examples of current research to examine how patterns of susceptibility or differential exposure may be viewed and understood at the level of the individual or the molecular level.

The workshop was held in May 1997. (See Appendixes B and D for a summary of the presentations and the agenda of the meeting.) It was designed to answer the three questions that comprised the charge to the group:

1. What areas within the existing portfolios of the sponsors are likely to yield information appropriate to this topic? What are the gaps in knowledge that warrant future research?

2. Are there research strategies and priorities for addressing the gaps in knowledge?

TABLE 2 NRC Reports with Information and Recommendations Related to Environmental Health Research and Training

Reports	Year
Biologic Markers in Pulmonary Toxicology	1989
Biologic Markers in Reproductive Toxicology	1989
Meeting Physicians' Needs for Medical Information on Occupation and Environments	1990
Addressing the Physician Shortage in Occupational and Environmental Medicine	1991
Animals as Sentinels of Environmental Health Hazards	1991
Environmental Epidemiology: Vol. 1, Public Health and Hazardous Wastes	1991
Opportunities in Applied Environmental Research and Development	1991
Biologic Markers in Immunotoxicology	1992
Environmental Neurotoxicology	1992
Indoor Allergens: Addressing and Controlling Adverse Health Effects	1993
Issues in Risk Assessment	1993
Measuring Lead Exposure in Infants, Children, and Other Sensitive Populations	1993
Monitoring Human Tissues for Toxic Substances	1993
Health Data in the Information Age: Use, Disclosure, and Privacy	1994
Biologic Markers in Urinary Toxicology	1995
Decision Making for Regulating Chemicals in the Environment	1995
Environmental Medicine: Integrating a Missing Element into Medical Education	1995
Nursing Health and the Environment	1995
Carcinogens and Anticarcinogens in the Human Diet: A Comparison of Naturally Occurring and Synthetic Substances	1996
Linking Science and Technology to Society's Environmental Goals	1996
Performance Monitoring to Improve Community Health	1996
The Environment and the Human Future	1996

3. What other strategies, including interagency coordination, might improve the prospect of developing knowledge that will identify gender differences in susceptibility to environmental factors?

The workshop was composed of two panels (see Appendix D). The first panel examined the overall issue of patterns of exposure among women. Presentations included issues related to environmental exposure in the workplace, environmental exposure and nutrition, and multiple environmental exposures over a woman's lifespan. The second panel focused on patterns of susceptibility, with presentations on various clinical and basic research studies. Presentations included epidemiology, gender, and environmental influences on multiple sclerosis; estrogen receptor knockout mouse studies and implications for hormonal differences in susceptibility; gender differences in metabolism and susceptibility to environmental exposures; and gender differences in the occurrence of molecular markers of carcinogenesis. The final panel included speakers and invited participants who discussed how current information is applicable to the three questions that formed the task. Part of the discussion was utilized as an opportunity to examine issues and obstacles to understanding women's differential susceptibility: recognizing variables in research on gender or sex differences; data collection, utilization, and analysis; understanding the relationships among exposure, dose, and effect; and the role of social factors in contributing to health and disease. The final portion of the discussion focused on a review of current federal efforts and resources and the creation of newer opportunities for collaboration among all federal agencies. Subsequent to the workshop, the committee met to discuss the results of the meeting and to outline areas for recommendations and priorities.

This report provides highlights of the workshop and a summary analysis of the research portfolios provided to the committee by the sponsoring agencies. Therefore, the committee's conclusions and recommendations regarding future research and policy directions for this area are based on the review of all materials at the organizing meeting, the analysis of agency portfolios (Appendix A), and the presentations and discussion at the workshop (Appendix B).

SUMMARY OF DISCUSSIONS FROM THE WORKSHOP

The summaries of the formal presentations at the workshop are in Appendix B. Below is a summary of the discussion by the committee, panelists, and workshop participants; it is based on the presentations and subsequent questions raised by those attending the workshop. The group was asked to identify issues related to the three parts of the statement of task and to suggest strategies for interagency coordination.

Discussion Points from the Presentations

One important issue is that of a woman's multiple exposures: that is, in the workplace as well as in the home, and during childrearing and caregiving for elders. For this reason, many participants suggested that exposure histories should be designed to collect more information about women's total experiences and exposures at different points in the lifespan. The group noted that there is growing scientific acceptance of the notion of differential susceptibility, but regulatory standards are, for the most part, still based on data averaged from male populations or experimental animals.

Weight gain and loss is another important issue. Women experience more cycles of fat gain and loss because of dieting behavior and also because of natural phenomena such as pregnancy. If toxicants stored in fat tissue are mobilized during these periods, this could be a significant behavioral/nutritional factor with differential impacts based on gender. Similarly, questions regarding neuroendocrine factors and stress deserve further attention. Regulatory changes may also have an impact on exposures for both men and women. For example, the Delaney Clause, a 1958 amendment to section 409 of the Federal Food, Drug, and Cosmetic Act of 1938, established a zero tolerance for pesticide residues and other known carcinogens in processed food. However, that clause is no longer in effect; therefore, there may be reason to be concerned about the level of toxicants found in food.

Research on "gender differences" and "sex differences" is often conducted and referred to as though they are the same; however, "sex differences" often serves as a proxy for cultural and socioeconomic variables that have little to do with biological differences between men and women. The group contended that the scientific community must do a better job of identifying biological differences in susceptibility, on the one hand, and nonbiological variables, on the other.

These other points were also raised during the workshop discussion:

- the rising proportion of women who are postmenopausal;
- the need to look at the role that infectious agents, emerging infections, repeated exposure to childhood infections, and differential exposures among childcare workers (the majority of whom are women) have on differential health outcomes between populations of men and women;
- ultraviolet radiation (men are more likely to work outdoors, but women are more likely to sunbathe);
- dietary and environmental estrogens as variables of the hormonal cycle; also, changes in dietary behavior that increase exposure to phytoestrogens;
- methods for detoxifying the body, such as chelation agents to bind and remove a toxicant, particularly if the detoxifying process makes once-stored agents more bioavailable;
- the role that race, ethnicity, and culture may have in establishing differential exposures between men and women of a given subpopulation or between women of different racial, ethnic, or cultural groups;
- accessibility of the raw data from the National Health and Nutrition Examination Survey-III (NHANES-III); these data recently have been made available to researchers.

There was a striking contrast between how the first panel addressed the issue of gender differences, in terms of global, societal, and cultural issues, and the second panel's approach, which focused on biological mechanisms. The group suggested that multidisciplinary research that included consideration of social and cultural factors was needed. Clearly one priority that emerged was the question of multiple or combinational exposures; another was the search for linkages between animal studies and clinical studies. However, at least one participant questioned whether the kind of multidisciplinary studies that would be needed are actually possible, given the fact that researchers from various disciplines would be looking at different points on the causal pathway, measuring with different tools, and obtaining different results.

The group concluded that it is important to gain a better understanding of basic biological mechanisms of pathways leading from exposures to health effects. They also indicated it was important to understand exposures, especially multiple exposures. There was general agreement that the potential exists for more synergy between basic scientists and epidemiologists than either group currently realizes. It was suggested that basic researchers could provide guidance on how to stratify an epidemiological analysis based on biologic factors, and epidemiologists can identify cohorts with specific exposures for further research on such issues as genetic variability. Issues suggested for further understanding include ways to shorten the time lag between basic discoveries, the applicability of laboratory findings in the clinical or epidemiological setting, and the availability of cheap, accurate diagnostic tools for monitoring exposures. In addition, the group pointed out that one desired result of exposure studies is to find ways to prevent or reduce those exposures.

As acute exposures with marked effects are eliminated, the chronic low-level exposures to other diverse factors may become more important. This suggests that nutritional interventions are also worthy of attention, particularly in view of the rising level of obesity in the United States. The group expressed the need for a systemic approach that looks at all these factors in women.

Several participants pointed to the need to adequately fund the fourth iteration of the National Health and Nutrition Examination Survey (NHANES-IV), to be conducted by the National Center for Health Statistics (NCHS) of the CDC. That is an efficient way to achieve comprehensive answers to some of the dietary questions. Money for planning NHANES-IV was known to be available—in fact, that planning is more or less complete—but no money is currently available for follow-up studies on NHANES-III. It was suggested that part of the reason is that follow-up studies would be based on secondary data, and the general opinion was that follow-up proposals do not fare well in the NIH peer review process. The group suggested that there is a need for coordination among agencies that fund such studies as well as among data collection systems generally. Several participants asked for improvement in the NHANES-IV measures of socioeconomic status (SES), occupational data, and environmental exposure measures. Others felt that researchers should be encouraged to submit proposals for secondary analyses of existing NHANES data that include testable hypotheses.

Opportunities for Agency Collaboration

A part of the discussion on existing federal resources and opportunities for interagency collaboration focused on several repositories of tissue and serum samples from earlier NHANES studies, including the availability of anonymous DNA samples from NHANES-III. The group noted that while NCHS had taken some steps to solicit the scientific community's input on the content of NHANES-IV, there was a need for additional input in the design of NHANES-V, perhaps through the mechanism of a broader advisory group.

Participants suggested that federal agencies should fund investigators to conduct a variety of secondary analyses that would integrate critical questions concerning gender and susceptibility, utilizing existing data sources; but they also wondered about the source of funding for such integrative studies.

A suggestion was made that one agency should centralize the resources that could be used by the entire research community. The National Center for Research Resources at NIH might be a candidate for this role. Opportunities for public-private cooperation to fund research that is too expensive for either sector alone were suggested as another possible option. Industry consortia have collaborated with EPA in this way, and both EPA and USDA have done cooperative research and development agreements with industry, university, and nonprofit researchers. Participants pointed out that CDC and NIH are perceived to be open to these cooperative efforts; the National Institute of Occupational Safety and Health (NIOSH) was highlighted as having industry staff working in its labs.

In areas where regulation would be a principal outcome, it was viewed as desirable to have a protocol workshop that would design a clinical or epidemiological study acceptable to all stakeholders. Another initiative cited was the development of core components on exposure and diet that were developed for the National Action Plan on Breast Cancer. The group suggested that consensus and protocol workshops are more common in science than in regulation; they emphasized that science should inform regulation.

In concluding the discussion, participants suggested that, in these and other areas, "environment" should be taken to mean more than just chemicals. Biomechanical stress, noise, and even violence in the workplace should also be considered as factors that belong on the list of environmental priorities.

PRIORITY ASSESSMENT: CONCLUSIONS AND RECOMMENDATIONS

Conclusions

Throughout the course of this activity, one area of confusion centered around the definition of terms. From discussions with the sponsors to discussions among the committee and the invited workshop speakers, it became clear that key terms needed to be defined. For the purpose of promoting useful interagency research strategies *additional factors should be included to capture aspects that are related to differences in susceptibility to environmental factors* (Table 3). Environmental exposures often implicate biological, chemical, and physical agents. However, in addressing the issue of gender differences in susceptibility to environmental exposures, the inclusion of nutritional, ergonomic (both biomechanical and psychosocial), and behavioral factors as well as specific places such as the workplace and the home in the concept of "environmental" provides a broader base for this type of research. Similarly, the definition of "susceptibility" could include all adverse outcomes except those that are directly related to hypersensitivity or allergy. Finally, the definition of "gender differences" can include aspects that are sophisticated enough to separate genetic and physiobiological differences between men and women from differences in environmental exposure, which in many cases result from the independent and interactive effects of socioeconomic status, employment patterns, and family, role and cultural experiences—sometimes called inherent vs. extrinsic differences.

The discussions involving speakers and workshop participants provided an important step in understanding how to (1) broaden the scope of review on gender differences and susceptibility, (2) identify areas in which there is or is not consensus, and (3) identify issues that need further attention. So valuable was this cross-agency and cross-disciplinary discourse that the committee concluded that *there is a need for such a workshop at regular intervals* in order to monitor progress and refine priorities for research.

TABLE 3 Additional Factors to Include in Broader Definitions

Term	Definition	Additional Factors
Environment	All chemical, physical, and biological features of the Earth that can affect or be affected by human activities.[a]	Include influences of specific factors such as nutrition, behavior, and ergonomics (both biomechanical and psychosocial) and specific places, such as the workplace and the home.
Susceptibility	The state of being readily affected or acted upon by the environment. Impact depends on exposure and the capability to respond.	Includes damage to all body systems, including reproductive organs, and offspring. Susceptibility also results from an inability to be removed from a toxic environment or to initiate primary prevention.
Differences	*Sex Differences* are primarily determined by biological factors, such as sex steroid hormone metabolism, anatomy, immunologic function, and genetic influences.[b]	*Gender Differences* include external influences from a variety of nonbiological factors, such as psychological development, sociocultural environment, and economic status.

[a]NRC, 1996.
[b]Ness and Kuller, 1997.

Preliminary review of the research portfolios provided by the sponsoring agencies suggests that they were developed on the basis of the standard definitions. In light of the additional factors identified by the committee, the research portfolios provided by the federal agencies comprise no more than a small subset of activities that the committee believes should be pursued. Although they appear to have been compiled according to divergent criteria, most of the enumerated projects were focused on *chemical* exposures and *reproductive system* outcomes.

The committee urges the agencies to *expand the review of their research portfolios, using the additional factors to augment the definitions*, in order to identify all possible influences upon women's health in an effort to prevent adverse health outcomes. Following this suggested expansion, an assessment of the relevant research portfolios should be conducted.[2] The goal of a reclassification effort based on this broadened approach of environmental factors is to develop a common database that will permit the agencies collectively to look at the adequacy of their funding in areas of greatest importance and to design interagency strategies and partnerships.

RECOMMENDATIONS

Research Recommendations

It became clear in the course of this activity that research on women's differences in susceptibility to environmental factors fell into two distinct categories: (1) differences in opportunities for men and women to be exposed to environmental factors, which are often culturally determined, and (2) genetic and physiological differences between men and women, which are biological. While these categories are useful for organizing research priorities, there is also a need for research that will bridge the gap between these two aspects of human life.

I. Priorities with Regard to Research on Exposure

1. **Research should be based on a broader inclusion of factors in the definition of "environmental exposure."**

In addressing gender differences in environmental exposures, it is important to broaden the inclusion of factors in the definition of "environmental" to include not only chemicals, physical agents, and pathogens but also nutritional,

---

[2]Subsequent to the workshop, the committee was made aware of a listing of federal projects related to women and environmental factors, developed by the Office on Women's Health, of the Department of Health and Human Services. However, this listing did not specifically highlight research designated in this area.

ergonomic (both biomechanical and psychosocial), and behavioral factors. In developing areas for research on the human health consequences of exposure to harmful physical and chemical agents in the environment, the relevant domains that should be evaluated are occupational and nonoccupational exposures, residence, physiological parameters, physical activities, and nutrition and diet. In the development of behavioral and medical interventions or the initiation of interagency research strategies and initiatives, special consideration must be given to gender-specific outcomes determined by occupation, behavior, lifestyle, hobbies, reproductive status, and physical activities. Examples of these various factors include but are not limited to:

- physical agents (e.g., noise, heat, vibration, ionizing radiation, pressure)
- pathogens (e.g., tuberculosis, HIV, cold viruses, *Candida*)
- nutrition (e.g., calorie intake, vitamin intake, ratio of fat in diet)
- biomechanical (e.g., sustained awkward postures, frequent repetitive movements, work requiring exceptional force and heavy lifting)
- psychosocial (e.g., job stress, level of demands from and control over work and life, amount of social support)
- behavioral (e.g., substance abuse, weight fluctuation, smoking).

2. **Population-based studies should include more complete and meaningful occupational data to help develop more accurate information on exposure.**

Many studies that are population-based, those investigating specific hypotheses (e.g., Framingham Heart Study, Nurses Health Study, etc.), as well as those designed to estimate population health status (e.g., NHANES) provide an excellent opportunity to gather data on exposure and susceptibility in both men and women. These studies are very useful for these purposes and represent unique resources for data collection. However, in most instances the data do not reflect the full range of potential exposures. This data gap is exemplified by deficient occupational data. A simple category of "occupation," for example, does not elicit useful information for a field like nursing, in which some workers are exposed to highly toxic chemotherapeutic drugs, while others are exposed to ionizing radiation, HIV and other infectious agents, ergonomic stress and anesthetic gases, violence, or perhaps a range of pediatric infections. Work histories are difficult and time-consuming to collect; hence, some compromise is needed to improve the "occupation" category. Methods for limited occupational history collection in some newer studies need to be validated.

Because national, population-based surveys are so valuable and because they are not often conducted, it is therefore paramount that the data collected be relevant and useful. The committee learned that planning stages for NHANES-IV are nearing completion. The possible inclusion of many of these questions and factors would help collect comprehensive information on a variety of factors that modify health outcomes, such as occupation. The NCHS should seek guidance from a broad research community to determine how best to record relevant occupational information in national surveys. Similarly, federally funded, population-based studies should be encouraged to improve exposure information collected by survey instruments.

3. Occupational exposure studies should adequately characterize and account for the full range of multiple exposures.

To explore the role of multiple exposures adequately, it is necessary to look at combinations and integrative effects over the full workday, the full working career, and the full lifetime. Many working men and women have a second (or even a third) job, each of which may have its own distinct set of exposures. Additionally, most women go home to "another shift" in which they have additional exposures, such as household chemicals and the emotional demands of childrearing (Broersen, et al., 1996). The combined impact of the demands of work and home for women was illustrated in a study of Swedish Volvo factory workers. This study showed that men and women had the same level of stress on the job; but when they went home at the end of the day, the men's stress levels went down, while the women's went up (Frankenhaeuser, 1989). Other studies highlight the need to study interaction and combined exposures between home and work life (van Dormolen, et al., 1990). The issue of multiple exposures to men and women is a critical and complex problem for population studies. Studies that examine effects of long-term exposures in the working population and susceptible groups (e.g., older workers, partially disabled workers) deserve special attention.

4. Research should examine gender differences in susceptibility to environmental factors over the entire lifespan as well as during critical exposure periods such as fetal development.

Presentations during the workshop demonstrated that both susceptibility and severity could vary over the lifespan. For example, early pregnancy is an obvious period of susceptibility of the fetus to teratogens; exposure of the growing child is also poorly understood as to the long-term health effects later in life; place of residence prior to age 15 is a major factor in multiple sclerosis (Kurtzke and Page, 1997); and smoking before 16 is a major factor in lung and breast cancer (Devesa, et al., 1995). Adolescence, in general, is poorly understood with regard to susceptibility to environmental exposure. There is evidence from both human and animal studies that initiation of mammary carcinogenesis originates in undifferentiated structures of the mammary gland during early adulthood. This model has been extensively developed by Russo and Russo (1997). Susceptibility before and after menopause is emerging as an important issue, as the number of women over 55 continues to increase. These and many other critical periods in the lifespan need further research to link the impact of exposures upon health outcomes.

5. **The development and use of appropriate animal models is encouraged.**

Some animal studies have examined differences in physiological response between younger and older animals. It is worth noting that a national resource already exists in this area: a colony of elderly rats supported by the National Institute on Aging. Although these rats are useful for many types of studies, they are hard to use for toxicology studies. The most direct animal model for human conditions may be nonhuman primates. As recently reported, a few federally supported facilities maintain large and expensive colonies of aging chimpanzees (Roush, 1997); utilization of baboons for relevant studies on aging may be feasible. Feline, canine, and porcine models are also useful for research on menopause-related questions. Further research is needed to develop more accurate animal models for a variety of other measurers of susceptibility studies.

6. **Studies are needed that identify the cultural and historical factors that account for the distribution of exposures between men and women.**

The issue of susceptibility by gender, masking what is really different exposure by gender, needs to be clarified. In other words, historical and cultural factors may have accounted for differential health outcomes. As such, the impacts of changes in our societal norm that may or may not have long-term health consequences need to be studied. Recent changes in our society include, for example, the influx of women in the workforce in all occupations and at all levels. In like manner, men are participating in homemaking and childrearing duties. Among the additional examples of changes in our society values that may impact on health outcomes are the growing numbers of grade school children and teenagers who smoke (IOM, 1994a) and the divergence in age of first pregnancies (increasing numbers of teenage mothers and increasing numbers of women who have their first child after age 30). It would be useful for social scientists to unravel the impacts and implications of these changes on health outcomes. Equally useful would be investigations on whether men and women respond differently to the same exposures or whether they actually have different exposures. The result might be to identify a kind of "socioeconomic marker" of susceptibility, similar to the genetic markers discussed in 10, below.

II. Priorities with Regard to Basic Research

7. **Basic research on gender and susceptibility to environmental factors should focus on (1) the biological basis for differences and (2) the contribution of environmental factors to the risk of disease from the same exposure.**

Mechanistic studies of hormone-dependent and hormone-independent processes or pathways should be investigated, beginning with animal/cellular models and followed by clinical research that translates the molecular-level information into the impact on human health. Here, too, however, the broader inclusion of additional factors in definitions of "exposure" and "susceptibility" comes into play. For example, stress was cited as a complicating factor in several presentations, and there have been several studies of the effect of stress on epinephrine levels in men (see Appendix B). Yet, relatively little is known about different responses and effects in women. It is known, however, that cortical responses modulate the immune response and that women's greater immune response contributes to differences in the appearance and severity of autoimmune disease. These types of examples suggest that the result may be differences in the prevalence of certain diseases (e.g., multiple sclerosis, lupus) between men and women and among women of different races and ethnicities (e.g., lupus is more prevalent among African Americans). However, differences in prevalence may not always be a predictor of health effect. For example, although fewer men than women contract multiple sclerosis, the disease is more severe in men than women, and the 10-year mortality rate is higher.

It may be that gender differences in neuroendocrine or immune responses, as well as differential exposures, will explain some of the gender differences in chronic diseases such as diabetes. Differential susceptibilities related to gender may ultimately prove important in understanding poorly defined syndromes, such as multiple chemical sensitivity. Consequently, research should address all these dimensions of physiology—hormonal modulations as well as neuroendocrine response—and the changes that men and women undergo over their lifespans.

8. **Priority should be given to studies of human diseases that are manifested differently in men and women or in which gender modulates susceptibility to environmental factors.**

Gender plays an obvious role in susceptibility to reproductive tract disorders, which deserve continued attention among both males and females. But gender also appears to play a significant role in modulating susceptibility to nonreproductive tract disorders. One obvious example is the role of estrogen in women's susceptibility to lung cancer resulting from exposure to cigarette smoke or some compromise of autoimmunity. On the other hand, estrogen can protect against bone loss, loss of vascular function, and possibly brain degeneration. Therefore, losses or supplements of this hormone may have profound gender-specific impacts. In some cases these differences appear to protect women, as in gastric ulcers and certain infectious diseases. The genetic and physiological mechanisms underlying these differences need to be studied and understood in order to identify effective procedures for prevention, intervention, or therapy.

Other examples of major public health problems which are manifested differently in men and women, or where environmental factors are manifested differently in women, include heart disease, pulmonary disease, autoimmune disorders, mental illness, and arthritis. This list of health problems is far from complete, nor is it currently possible to set priorities among them. The identification of all such diseases and the setting of actual priorities among them should be part of any effort to set overall research priorities.

9. **Research should examine the impact of metabolic differences between men and women, as well as neuroendocrine, immune, and hormonal differences.**

There are differences in metabolic processes between genders. As indicated in the workshop summary (Appendix B), the herbicide 2,3,7,8-tetrachlorodibenzo-p-dioxin (TCDD) induced liver cancer in female Sprague-Dawley rats, but not in males. Hormonal components are involved in this gender difference. On the other hand, there was a significant overlap and interaction between metabolism and susceptibility genes in the study of markers of exposure and susceptibility to carcinogenesis. However, we do not know if it is possible to extrapolate the effect of this toxicant upon humans. The role of gender differences in these interactions has not been clearly identified, and the whole field deserves further study.

10. **Research should seek to characterize genetic markers of susceptibility.**

Genetic polymorphisms and mutations play a potential role in ethnic as well as gender differences with regard to susceptibility to environmental factors. Research that identifies specific genes or combinations of genes that are reliable predictors of susceptibility could produce enormous benefits for both prevention and diagnosis. Once identified, the genes could be studied in cellular or animal systems in order to study mechanisms and effects. A recent new initiative by NIEHS to collect susceptibility genes for large-scale studies on how these genes vary from person to person is a major step in this area (NIEHS, 1997). Information that is developed from this project will contribute greatly toward knowledge of the genetics contribution to susceptibility to environmental factors. However, as in any research on human subjects, great emphasis must be placed on the protection and privacy of the subjects. There are many instances in which existing databases may be of use to researchers without jeopardizing the privacy of individuals. For a fuller discussion of issues related to privacy, see the IOM report, *Assessing Genetic Risks* (IOM, 1994b).

11. **There is a critical need for translational research to bridge the gaps among cellular, animal, and human systems.**

When epidemiologists, for example, identify a relationship between risk factors and disease that suggests gender-specific differential susceptibility, researchers then need to study the basis of that relationship at the molecular and cellular level. In addition, there is much to be learned from animal models, not only from the knockout mice (as described in the workshop summary) but also from gene insertion and substitution techniques (e.g., so-called "knock in" mice and "hit and run" experiments).

At the same time, however, new techniques are needed to validate animal models of human disorders; assistance in translating those validation techniques into simple tests that can be used in the field are equally necessary. For these reasons, basic researchers should look for the broader implications of the mechanistic research conducted at the molecular, cellular, and animal levels. In particular, sponsoring agencies should encourage the development of animal models that are directly relevant to (1) gender differences in susceptibility and severity and (2) human exposures and diseases.

III. Policy Recommendations

12. An annual workshop should be held to encourage and promote opportunities for interagency collaboration.

Research priorities for improving the understanding of gender susceptibility to environmental factors occur in at least three different major areas of research: worker health and safety, women's health research, and environmental health. Each area is supported by a separate funding stream. Synergy can be achieved through a careful review of both the established priorities and missions of the agencies with the cross-disciplinary research needs to identify areas of priority. One way of identifying those areas of natural congruence would be through workshops jointly sponsored by relevant agencies at regular intervals. The workshops would also help monitor progress and refine priorities for long-term studies in this area. The joint sponsorship of this IOM review was an encouraging sign. Presentations by the sponsoring agencies' representatives indicated that other jointly funded efforts are occurring also. The committee encourages these partnerships and believes that more interagency cooperation would advance the research base in this area. Therefore, the committee believes that the proposal for an annual workshop would significantly encourage the development of joint activities.

Suggestions for agenda topics for such a workshop abound. One would be to develop common definitions and categories to guide a broader review of current research activities; others would be to develop a government-wide research priority list (subject to annual revision) and invite proposals for an interagency initiative. The committee suggests neither a particular administrative structure nor a particular lead agency; rather, it encourages agencies to cooperate in setting priorities, funding extramural research, and/or conducting intramural research with multidisciplinary interests.

13. Agencies should work together to make necessary institutional changes.

Many of the research priorities described are interdisciplinary in nature and will require interdisciplinary peer review. Such interdisciplinary review groups are more challenging to conduct; hence, all sponsoring agencies should work together to meet the peer review challenge.

Interagency cooperation and collaboration are, however, not limited to project review. Mechanisms and programs are already in place and attempt this kind of "cross-boundary" coordination; they may provide models for future collaboration. For example, an intra-NIH advisory committee, convened by the Office of Research on Women's Health, involves all the NIH institutes. Members of the Interagency Task Force on Women and the Environment, established by the Department of Health and Human Services (HHS), come from Cabinet-level departments. Experience indicates that these groups work best when they have high-level and consistent participation.

14. Current sponsors should make every effort to expand the roster of agencies conducting or funding research on gender and the environment.

The scope of this project was limited to the civilian agencies that sponsored it. Nonetheless, many, if not all, research-based federal agencies have missions and programs that support various aspects of women's health and environmental health. The Department of Defense (DOD), for example, has a number of research programs that can contribute to the development of new knowledge regarding gender differences and differential susceptibility. Servicewomen are increasingly exposed to the same environmental risks as their male counterparts. This type of research could become a high priority in DOD's long-term planning. In addition, DOD's classification of jobs through its Military Occupational Specialties system could be the basis for a more specific occupational analysis than is available in most other civilian-based data sets. DOD already supports some analysis in this area; it might prove fruitful to hold discussions between DOD and NIOSH or NIEHS with regard to joint support of work-related research. DOD already has women's health programs and, therefore, has a built-in interest in gender-specific research. More importantly, DOD has a peer review system already in place; this would facilitate review of meritorious research proposals on topics of interest to this project. Another equally important partner is the Department of Veterans Affairs (VA) which will have a growing interest in gender differences and differential susceptibility as more former servicewomen become eligible for medical care in VA hospitals.

15. Sufficient interest and opportunity exist for agencies to invest in prospective research projects that focus on both gender differences and the environment. These investments should be

flexible with regard to funding mechanisms and should provide continuity for long-term investigations.

Many of the recommended research objectives could be accomplished by giving the combined issues of gender and environment higher priority within agency budgets. However, agencies should also look for innovative opportunities to match their funding mechanisms with other types of research support. Traditional investigatior-initiated R01 grants at NIH would suffice for many research proposals; but the use of other mechanisms such as project grants, core support, subcontracts, and cooperative agreements should not be ignored. These other mechanisms can be tailored to specific agency needs and priorities and can provide support over a relatively long period of time. This long-term support is vital for continued surveillance of human populations and multiyear funding of laboratory and animal resources. In such cases, continuity of funding, possibly through bridging mechanisms, can be extremely important. Funding versatility and continuity of support are vital for providing the necessary infrastructure to encourage research on gender differences and susceptibility.

16. Opportunities for cofunding and for public/private cooperation with university, nonprofit, and industry groups should be sought.

While the preeminent federal role is to create new knowledge, the line between basic and applied research is sometimes blurred. Within this gray area, however, cooperative relationships with the private sector can produce great mutual benefits (NAS/NAE/IOM, 1995). Gender differences in susceptibility are not only important to many government agencies but also to many private organizations willing to form research and funding partnerships. Some of these private-sector groups include foundations, universities and university consortia, labor unions, and industry (particularly drug companies). Partnerships with private industry, such as pharmaceutical companies, for research in gender differences and susceptibility could produce data and information that may have both long-term biomedical significance and short-term value for product development and marketing. A recent example of such a partnership occurred in the Women's Health Initiative (WHI).[3] A $9 million "add-on" study, completely funded by the private sector, now conducts research on the effect of hormone replacement therapy on cognition and Alzheimer's disease. Because the WHI involves a sizable population of women of all races and lifestyles, the data de-

---

[3]The Women's Health Initiative, an activity centered in NIH, is focusing on the major causes of death, disability, and frailty in postmenopausal women. The overall goal of the WHI is to reduce coronary heart disease, breast and colorectal cancer, and osteoporotic fractures among postmenopausal women through prevention/intervention strategies and risk factor identification. WHI is a 15-year effort that is budgeted for $625 million.

rived from this "add-on" study will aid the WHI, the private-sector sponsor, and the women who participate.

In the arena of innovative cooperative efforts, more attention should be given to partnerships with nontraditional partners, such as nongovernmental organizations (NGOs). NGOs are the site for some of the most innovative work that is being done on linking gender and environment, including a focus on women's health outcomes.

17. **Strategies for utilizing national health surveys and databases should be developed. The broader public health community should be encouraged to find ways to improve and broaden such utilization in the future.**

Large-scale health surveys and their databases, particularly longitudinal and cross-sectional data from study populations, contain potentially valuable information that could be used to study environmental issues. CDC alone has data from its National Center for Health Statistics (NCHS), National Health and Nutrition Examination Survey (NHANES), National Health Interview Survey, National Ambulatory Medical Care Survey, and National Disability Study. Other studies that examine specific cohorts include the ongoing Nurses Health Study and the Six City Study. These surveys and databases are well known to a narrow constituency but not so well known to the broader scientific community. Given the significant investment that has already been made in collecting this information, it would be very cost-efficient to invest a bit more for further analyses.

Access, however, remains a critical barrier to greater utilization of these resources. Although CDC is increasing its efforts to make data more accessible by providing tapes and CD-ROMs, the results are still not user-friendly, nor are the data easily exported into common statistical data analysis software. All federal agencies should be encouraged to develop pathways for easier public access to the information.

Another barrier to access is the issue of confidentiality of stored tissue samples. Many states store DNA samples for long periods after their initial use. In its report, *Assessing Genetic Risks* (IOM, 1994b) the committee stated that:

> Later access to DNA samples or to the profiles for other purposes should be permitted only when . . . b) the data are to be anonymously studied. . . . In general, regardless of the purpose for which it was compiled, this information should be accorded at least the confidentiality that is accorded to medical records.

This committee suggests that NCHS and CDC take the lead in conducting a strategic review of their current portfolio of population-based health surveys and databases and develop guidelines for public use of the data for purposes other than national extrapolations. A general review of these data resources and the development of strategies to fully use them would comprise an appropriate agenda for an interagency workshop (see III-1 above). Some specific issues to be addressed at such a workshop might include privacy, confidentiality, informed consent, ownership of samples and data, and the identification of new uses and users.

Once the current survey portfolio is known, a review of the types of data it contains would be the next step. In some cases the data may have limited application, because they are incomplete or noncomparable or because they were gathered with survey methodologies that are now outdated. After identifying the strengths and weaknesses of the current portfolio of surveys, agencies should ask the broader public health community to help find ways to improve and broaden their usefulness in the future. The goal should be to expand the utility of the information and broaden the community of users, without compromising the original public health purposes of the surveys. With advice from the user community, ways should be explored for improving and integrating these surveys and databases and for advertising their availability to a broader community of interested researchers.

The IOM report, *Toxicology and Environmental Health Information Resources* (IOM, 1997), makes similar recommendations. Many of that report's recommendations focus on the need to heighten awareness of health information resources, to analyze user profiles, and to simplify the navigation into and through the databases for toxicology and environmental health. These recommendations should be expanded and applied to broader fields of health data and information in order to assist the development of new knowledge for gender differences and susceptibility.

18. Strategies should be developed to identify, protect, and utilize other irreplaceable research resources.

DNA, serum, and tissue samples have been collected in conjunction with some of the health surveys described above. While these resources were collected for specific purposes, the committee suggests they may have broader uses. Geneticists, for example, would be interested in the large population of "normal" DNA contained in such collections. Such uses would need to be consistent with current ethical standards and laws governing confidentiality, informed consent, and privacy.

The National Cancer Institute supports a number of cancer registries whose resources are generally available to interested researchers. Additional resources of this sort are maintained by private nonprofit groups: the American Type Culture Collection, for example, collects human normal and tumor cell lines as well as microbial cell lines; the Human Biological Data Interchange collects longitudinal samples from families with autoimmune disorders.

An inventory of these resources would be useful and might reveal the need for (and sponsorship of) new components or new collections that would make these existing surveys and databases more useful and productive. A fuller description of the issues, problems, and features of successful resource sharing is contained in the IOM report on *Resource Sharing in Biomedical Research* (IOM, 1996b).

CLOSING REMARKS

A summary of the committee's recommended priorities that were discussed above is displayed in Table 4.

The committee believes that these recommendations, taken together, provide a good beginning for the identification of research priorities and interagency initiatives that are well within the mission of the various agencies involved. As additional resources become available, it is hoped that research priorities in this area can be implemented. Opportunities abound, and the new knowledge that can accrue will be valuable for all women throughout their lives.

TABLE 4 Summary of Recommended Priorities

Exposure Research	Basic Research	Policy
Broader Definition The definition of "environmental exposure" should include additional factors to capture information relevant to the unique susceptibility of women.	**Biological and Environmental Causes** Basic research should focus on the biological basis for gender differences and the contribution of environmental factors to the risk of disease from the same exposure.	**Annual Workshops** Sponsor agencies should hold annual workshops to identify and act on opportunities for interagency cooperation.
Occupational Data Elements Population-based studies should include more complete and meaningful occupational data as part of an effort to develop accurate information on exposure.	**Gender Differences and Diverse Outcomes** Priority should be given to studies of human diseases that present differently in men and women or in which gender is a modulating factor for susceptibility.	**Institutional Changes** Agencies should work together to make necessary institutional changes.
Multiple Exposure Data Occupational exposure studies should adequately characterize and account for the full range of multiple exposures.	**Metabolic and Hormonal Differences** Research should examine the significance of metabolic and hormonal differences between men and women.	**Participating Agencies** Current sponsors should make every effort to expand the roster of agencies conducting or funding research.
Lifespan and Critical Exposures Research should examine gender differences in susceptibility to environmental factors over the entire lifespan, as well as during critical exposure periods.	**Genetic Markers** Research should seek to characterize genetic markers of susceptibility.	**Long-Term Prospective Studies** Agencies should invest in prospective research projects that focus on both gender differences and environment. Investments should be flexible with regard to funding mechanisms and should provide continuity for long-term investigations.
Animal Models The development and use of appropriate animal models is encouraged.	**Translational Research** Translational research is needed to bridge the gaps among cellular, animal, and human systems.	**Public/Private Cofunding** Opportunities for cofunding and for public/private cooperation with university, nonprofit, and industry groups should be sought.
Cultural and Historical Factors Cultural and historical factors that account for the distribution of exposures between men and women should be identified.		**Improve Access and Content of Data** Strategies for utilizing national health surveys and data sets need to be developed. Input from the research community is needed to shape broader application.
		Irreplaceable Resources Strategies to identify, protect, and exploit other irreplaceable research resources also need to be developed.

A

Analysis of Agency Research Portfolios

INTRODUCTION

As one of its first tasks, the committee completed an analysis of the research project summaries provided by the National Institutes of Health, the Centers for Disease Control and Prevention, and the Environmental Protection Agency. The analysis was in two parts: an initial assortment into six categories of research being conducted by each agency, followed by a more subjective analysis of each research category across the agencies. These analyses were conducted after the assignment of every project to one of six descriptive categories. The categories and their definitions are as follows:

1. **Environmental exposures (EE)**, defined as studies or activities that examine specific agents in the environment and their impact on health.
2. **Endocrinology (END)**, defined as studies that examine the impact of environmental exposures on the hormonal/endocrine systems. This category also includes the impact of specific agents on the endocrine systems.
3. **Long-term chronic care (LTCC)**, defined as studies that assess the impact of long-term environmental exposures on specific diseases or disabilities.
4. **Molecular biology/basic science (MB/BS)**, defined as studies that focus on environmental exposures at the molecular level; includes animal studies.
5. **Epidemiology (EPI)**, defined as studies that examine the impact of environmental exposures on populations.
6. **Prevention (PREV)**, defined as studies or activities that have as their primary goal to prevent the adverse health consequences of environmental exposures in populations.

The purpose of this analysis was to obtain an overview of research areas supported or not supported by federal agencies. The results of this analysis were factored into the process of selection of topics for the workshop and the identification of issues for the priority recommendations.

Analysis of Agency Projects

The National Institutes of Health (NIH) provided 121 project summaries; the Centers for Disease Control and Prevention (CDC), 41 summaries; and the Environmental Protection Agency (EPA), 15 summaries for a total of 177 project summaries reviewed. The relative proportion of each category in each of the agencies is shown in Figure A-1. As expected, the types of projects vary by agency and are related to the mission of each.

Analysis of Portfolios by Subject Category

A second, subjective analysis was done based on sorting the projects into the six categories listed above. The results and observations by the committee are below.

Environmental Exposures

Areas highlighted in the portfolios. Research areas represented in the portfolios provided by the agencies include work associated with some of the epidemiological studies from family exposure research; men and women are included. In others, exposure measurements were combined with epidemiological studies as, for example, in the animal models of chemically induced endometriosis. Also, the portfolios contain studies on actual or theoretical biomarkers of exposure.

Gaps/missing areas of research. Few studies in this category deal with environmental exposures and women's health. None of the studies covers the entire lifespan of women: that is, environmental exposures in utero and in children, young women, women of childbearing age, and postmenopausal women. A corollary set of studies would be those that focused on exposures during critical times throughout the life span and during development. An area of increasing interest will be studies on frail elderly women in their 80s and 90s. Weight loss gradually occurs among women in their 80s and 90s, and loss of fat cells exposes their bodies to any toxicants stored in those cells. The same scenario is possible as bone mass is depleted.

Not many studies include environmental exposure measurements, especially in epidemiological studies. In most studies reviewed by the committee, exposure is considered as a variable but not as the main focus. Also, studies are needed that describe the nature of exposure settings (i.e., home, workplace, etc.) in order to develop new treatments or polices for intervention.

APPENDIX A

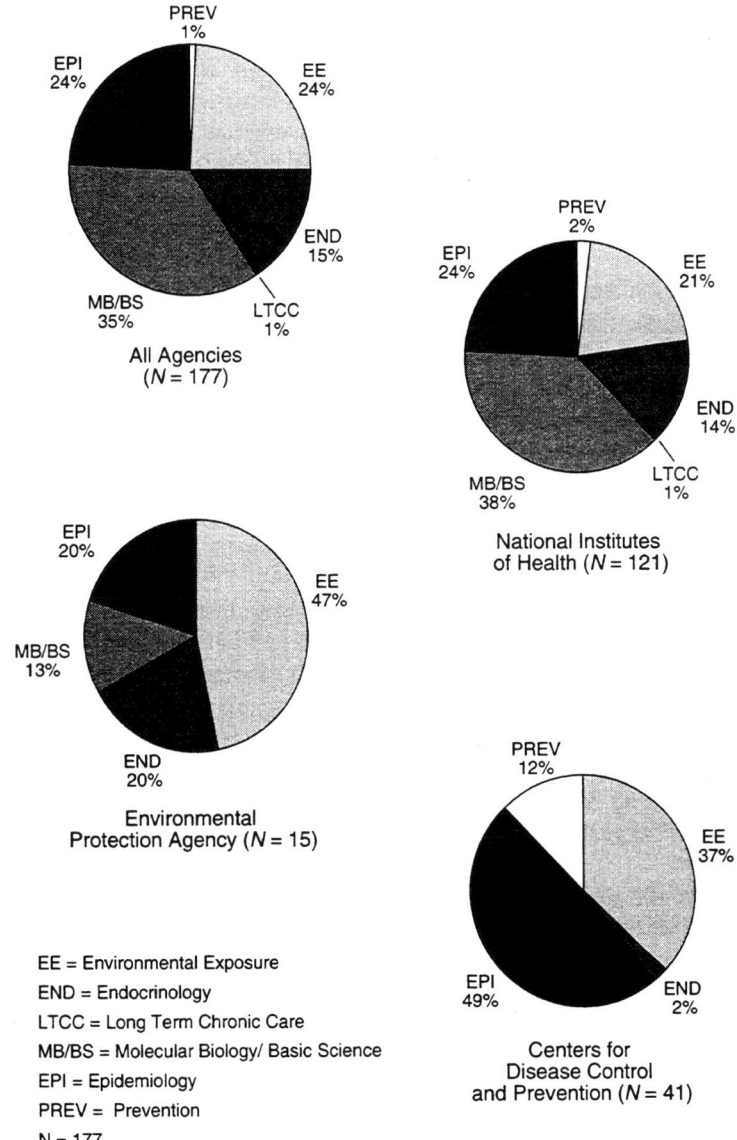

EE = Environmental Exposure
END = Endocrinology
LTCC = Long Term Chronic Care
MB/BS = Molecular Biology/ Basic Science
EPI = Epidemiology
PREV = Prevention
N = 177

FIGURE A-1 Percentage of research portfolio by category.

Endocrinology

Areas highlighted in the portfolios. This category also included the impact of specific agents on the endocrine system. Endocrine-related research in the agency portfolios included a disproportionate amount of research on breast cancer, breast development, and hormonal breast regulation. Pregnancy outcomes, placental effects, marker enzymes, genes, in utero exposures, and the effects on female offspring were also covered in a number of studies. The field of biomarkers is expanding, particularly as it involves drug-metabolizing enzymes. While the research studies on marker enzymes are deemed important by the committee, there is a need to focus research on the importance of several indicators, which, if they occur together, constitute a more significant clinical marker.

The committee was encouraged by the number of ongoing projects focused on exposures to the uterus and the endometrium. Uterine research is important, given the magnitude of the problem of endometriosis and the fact that animal models for the condition are difficult to obtain. (It was noted that the National Institute on Child Health and Human Development recently announced a request for proposals for endometriosis research.)

Projects focused on chemicals study the usual dioxins, PAHs, PCBs, and endocrine disrupters. A number of ongoing projects address the effects of these chemicals at critical times during a woman's development. The committee was impressed by the number and breadth of the projects and the number of chemicals being studied in relationship to the endocrine system.

Gaps/missing areas of research. The following areas were identified as deserving of future work:

- **Solvents**: Except for dry cleaning solvents and the usual top half-dozen or so major chemicals of concern, studies on solvents were not well represented in the portfolios.
- **Neuroendocrine research**: A focus on women, depression, and disease was not well represented.
- **Metals**: The spectrum of metals being studied needs to be broadened.
- **Immunopathology**: Of particular importance are diseases with an immunopathological aspect, such as scleroderma (also related to exposure to solvents).

Long-Term Chronic Care

Areas highlighted in the portfolios. While there was only one project in this category, the committee noted that other projects had aspects of this issue contained within them.

Gaps/missing areas of research. All future research in this area should address multiple age groups, unless there is a research issue that occurs during a critical age range.

Molecular Biology/Basic Science

Areas highlighted in the portfolios. The committee noted that basic research studies posed questions about toxicology; however, fewer toxicological studies were present that studied the effects of compounds on biological systems. This may be because studies on whole animals, exposed to teratogens, are difficult to interpret. The committee found that the portfolios in this category were heavily weighted towards research on breast cancer. They also noted a major emphasis on endocrine disrupters, disproportionate (in the committee's opinion) to the diseases caused by those compounds.

Gaps/missing areas of research. The committee noted that recent basic science research findings have expanded the questions posed that need to be answered through toxicologic studies. There are few examples where toxicologists designed mechanistically-oriented studies to provide a more detailed understanding of the effects on systems of environmental exposures. The committee identified several areas for future research. The first is for research on mechanistic approaches, which ask questions about pathways and targets for disruption, as opposed to the toxicants themselves. There is a need for greater interest in the underlying biology and in using compounds to ask questions about biology. Other future needs include a more comprehensive approach to understanding biological mechanisms and how toxicants alter biological functions. In summary, more research is needed which takes advantage of recent advances in molecular and cell biology to address how toxic compounds alter normal physiological processes, including gender-dependent targets.

The committee indicated that current compound-specific research has a somewhat narrow toxicological focus; the committee, however, advocates a broader biological approach. In addition, there is growing interest in the molecular imprinting process that may occur as a result of perinatal exposures. The committee believes this area deserves more thought and attention and considers toxicological questions to be tied to the basic biological questions. Therefore, a concerted effort to conduct research on the molecular imprinting process in perinatal exposures would be productive for the whole field.

Many other organs and organ systems, other than breast development and regulation, need attention. There are also many serious questions in breast research other than those concerned with cancer. The portfolios did not contain much information on the role of dietary effects, particularly regarding phytoestrogens.

The committee noted also many opportunities for the use of animal models that express human genes and those that control genes using human gene regulatory elements. Also, there was not a significant cohort of long-term molecular studies in animals. Such studies were not represented in the portfolio, even though they may provide the basis for answering important questions about toxicants in systems utilizing genetic technology.

Epidemiology

Areas highlighted in the portfolios. The committee commented that beyond the projects placed within the epidemiology category, there were an equal number of project summaries in other categories that had epidemiological components.

Gaps/missing areas of research. The committee noted that the portfolios contained very few projects concerned with environmental exposure. Much is known about the relationship of exposures to disease in women which could lead to prevention. Of critical importance would be new research to describe the nature of exposure settings, in order to develop appropriate interventions. The committee referred to the missing component as "hazard surveillance" or "hazard epidemiology." These type of studies may not be typical for biologically-based agencies, but are important links to biological measurements and analyses. Other missing aspects include differential risks of exposure according to women's work and women's lifestyles, whether or not they are unique to women.

Another area needing attention is a careful assessment of the impact of multiple exposures. Few studies focus on one substance; many focus on multiple substances without understanding the difficulty of such studies. Recent data suggest the need for a more careful look at the synergistic effect of multiple exposures, as it relates to organochlorines and estrogens.

Some windows of opportunity were noted, primarily by the CDC projects on the magnitude of risk. There are some pharmacokinetic, pharmacodynamic, and biological exposure models used in epidemiology that do not currently include gender. The committee suggested that the importance gender may play in these models is inadequately examined and deserves a more intense focus by investigators who choose to develop and examine these models. Statistical methods are another area in need of support in epidemiology, in general, and epidemiological studies of women's risks, in particular.

A series of data systems is needed to facilitate both analysis and surveillance. These systems could be used to identify risks, to track risks, and to track the effects of intervention on risks. Most of the NCHS data are used by NCHS descriptively. The committee views the NCHS database as an impressive resource that is not being fully utilized. More knowledge can be leveraged from these data because of the risk information that is collected. Beyond national databases, the committee observed that there are not many population-based studies, longitudinal studies, or cohort studies in the portfolio. The few that were highlighted were internal studies identified as part of the NIH portfolio.

While important work should continue on chemical toxicology and its relationship to women's health, other work needs to be done with a focus not on chemicals, but on psychosocial effects and risks (e.g., on cardiovascular disease). The committee pointed out that the effects of behavior on biology and disease induction are a complex but critical area for future research. Some studies of the cardiovascular demand/control model with a social support component suggest that the combination of demand, control, and support is more important than cigarette smoking in the etiology of cardiovascular disease. The committee urged researchers to look for important risk factors in cardiovascular disease, even though the technology of exposure assessment is complex and utilizes behavioral science measures. Job demand and job control in women versus men was also not among the studies reviewed. The committee noted that Europeans seem more interested in exploring these issues; but even the European studies are *controlling for* gender rather than *focusing on* gender.

Finally, the group noted that there is a need to look at environmental allergic disease and ascertain in what ways women are different from men, in order to develop hypotheses for intervention.

Prevention

Areas highlighted in the portfolios. The committee noted that the prevention portfolio was small; the bulk of it is in CDC. (The committee wondered whether a broader search of the NIH database would have identified any additional prevention projects).

The committee noted that the CDC portfolio generally focused on population-based models of prevention, where either public information or advertising was used in some way so that the media transmitted the information about prevention. Because the analysis of the EPA portfolio did not result in the identification of projects in this area, it was suggested that a definition of prevention linked to regulation might uncover additional EPA prevention research. Comments by EPA representatives indicated that only those projects with a strong research component or women's base were submitted, and that other activities were not included.

Gaps/missing areas of research. The committee suggested that there undoubtedly were more projects in EPA and NIH that benefit society as a whole and not just women. Therefore, the committee urges against the interpretation that there is a lack of prevention-related research in EPA or NIH. Of the other projects reviewed, only one project had risk assessment in the title or in the abstract; it raised the question of the agencies' view of prevention research and the operative definitions and models for prevention research across the agencies. Policy-related research, such as that informing changes in the use of toxic substances or uses of engineering controls, was not evident in the portfolios.

Nonetheless, the committee suggested that it was not surprised that the prevention area is relatively small. The committee endorsed the notion that rushing ahead with preventive interventions without fully understanding the biological effects of these interventions is not good. Prevention research should focus on individual behavioral change, an emphasis on regulation, and widespread dissemination of protective measures that people themselves can employ to limit exposure.

The committee noted that much of the ongoing preventive activities are not gender-specific; opportunities for examining gender-specific factors would be useful and would improve the potential for prevention. Such opportunities should include different gender-specific learning practices or behaviors. The committee also suggested taking advantage of the behavioral science literature that deals with preventive strategies.

CONCLUSIONS

In conclusion, the committee suggested that the overall research portfolio is not balanced between exposures and the presumed mechanisms that affect our health. This is particularly true for exposures relevant to women who work. As for epidemiology, some mainstream research issues are being addressed, but some critical issues are missing from the portfolio.

Discussions with the agency representatives revealed three factors that should also modify the interpretation of these analyses:

1. *Definitions of terms are not necessarily consistent across the agencies.* Prevention and preventive strategies for prevention, for example, may be categorized by one agency as prevention and by another as intervention. Intervention may occur with a specific drug, but it may also be preventive strategy.

2. *Other research, not particularly focused on women's health, may nevertheless be relevant.* This may be the cause of an underestimate of the complete portfolio that is directly applicable to the task.

3. *Research from other components of the agencies may not be represented in the project listings.* Because a limited number of NIH components are participating in this study, other relevant work in other components may also be relevant to the task.

These caveats, taken together, suggest that this analysis is, at best, an underestimate of the work that is being done and is a subset of research activities in the agencies. Nonetheless, a review of the projects identified thematic areas in which substantial research is being conducted, as well as areas in which there are gaps in research and opportunity. The committee suggested that the agencies review their research portfolios using identical definitions, expanded to include all possible adverse influences on women's health. Following this suggested expansion, an assessment of the relevant research portfolios should be carried out again.

B

Summary of Workshop Presentations

This section summarizes the presentations and discussions by speakers, committee, and participants at the IOM workshop. (See Appendix D for agenda and lists of speakers and participants.) This workshop was useful for understanding how to broaden the topic, identify areas in which there is or is not consensus, and spot issues that need further attention. The focus of the workshop was to address the three questions that comprised the study's statement of task:

1. What areas within the existing portfolio are likely to yield information appropriate to this topic? What are the knowledge gaps that warrant future research?

2. Are there research strategies and priorities for addressing the knowledge gaps?

3. What other strategies, including interagency coordination, might improve the prospect of developing knowledge that will identify gender differences in susceptibility to environmental factors?

In addition to the above three considerations, each panelist was asked to provide suggestions for further research and interagency collaboration.

The workshop was composed of two panels. The first panel examined the overall issue of patterns of environmental exposure among women. The second panel focused on patterns of susceptibility to environmental factors. Speakers on each panel also participated in the general discussion of how current information applies to the three questions above. The balance of the workshop was devoted to a general discussion of federal efforts and resources and the opportunities for collaboration among federal agencies.

PANEL I: PATTERNS OF EXPOSURE

Environmental Exposure in the Workplace[1]

Research on gender differences in susceptibility to environmental exposures in the workplace must answer a number of questions relating to physiological and hormonal differences, differences in susceptibility and deposition of toxicants, and to metabolic and genetic differences between men and women. These issues will be addressed by other speakers. However, research on environmental exposure and gender susceptibility also requires careful attention to issues of methodology and experimental design. For example, there are no standardized definitions of job titles or job content, making it difficult to compare the actual exposures or outcomes of different groups of workers, male and female. Failure to address these conceptual problems will lead to inaccurate data, misleading findings, and poor policy decisions.

Approximately 20 years ago, General Motors, among other employers, established a so-called "fetal protection policy" that simply excluded all fertile women from jobs that involved exposure to lead, regardless of the women's intentions with regard to childbearing. Nearly two decades later, the Supreme Court, in a 9-0 decision (*UAW v. Johnson Control*), ruled that employers could not establish such fetal protection policies, because there was abundant evidence that pregnant women were not the only group of workers vulnerable to the effects of lead. One of the lessons learned from this experience is that research which seeks to explore issues of differential vulnerability to toxins in the workplace should not be narrowly focused on those groups perceived as most vulnerable but should encompass all groups that are "at risk."

When research is carried out on a particular biological process, such as pregnancy, care should be taken to avoid over-extrapolation of the results to broad categories of workers, as this can lead to the development of poor social policy. Conversely, research which is grounded on clear definitions of the limitations to generalizability and on hypotheses that take into account the social and physical realities of the workplace and of employment can provide a sound scientific basis for prevention and control of environmental hazards in the workplace.

A major methodological problem in carrying out such research is that men and women do not encounter the same environmental exposures at work. For the most part, American women remain in "ghettoes of employment." Today there are more women in the workforce, but the traditional employment patterns of women have not changed significantly over the past 50 years (see Figure B-1). Twenty-one percent of all professional women are still engaged in teaching in secondary (i.e., below college-level) education. Women still dominate in

---

[1]This section is based on the presentation by Jeanne Stellman, Ph.D., associate professor of clinical public health at Columbia University.

primary education, government, and clerical employment. Men still dominate in industries with the greatest potential occupational exposure to hazards (e.g., construction, manufacturing).

Furthermore, when men and women are employed in the same sectors, the distribution of job titles is different. One example is the health care sector, where 72 percent of the workers are women, but this includes 94 percent of the registered nurses and only 20 percent of the physicians. And even when men and women have the same job title, women, in general, have different tasks than men. In one study of workers in the Canadian poultry industry, for example, women working within the processing sector had an elevated accident rate compared to the male workers. When the data were analyzed according to the actual tasks performed, and when the working conditions were surveyed, it was found that there were substantial differences in the nature of the work and its physical construction that could successfully explain a great deal of this reported rate difference.

For example, the height of the workstation, the design of protective equipment, the rate of work, the number of breaks, and the physical mobility around the plant placed women workers at a disadvantage. In practice, it is very difficult to find examples of work environments where men and women are exposed to the same environmental risks and factors. In "high-exposure" industries, furthermore, there have not been enough women employed, exposed to enough risks, and for enough time to draw statistically meaningful conclusions about differential susceptibility. Inconsistencies in employment statistics collected by various federal agencies create additional problems in data collection and analysis for occupational environmental research.

An even greater difficulty in research on gender and occupational exposure, however, is the complexity of the conceptual framework. In general, research has concentrated on the independent variables (the so-called "host factors") such as reproductive state, toxicological mechanisms, and individual metabolism. In general, the state-of-the-art of occupational health research is not advanced with regard to identifying the full range of relevant variables, such as socioeconomic status and lifestyle and defining how these variables interact. Significantly, these factors can be either independent or intervening variables, thus further complicating the conceptual framework.

Failure to take into account all these interactions in the conceptual framework can lead to incorrect conclusions and poor social policy that may ultimately be detrimental to the well-being of working women and men. Environmental health research often finds itself extrapolating from individual traits into population-based results and conclusions. It is important to remember that women and their exposures should be treated as separate variables; personal characteristics and environmental variables are the building blocks of any conceptual framework, and one cannot simply be considered a proxy for the other.

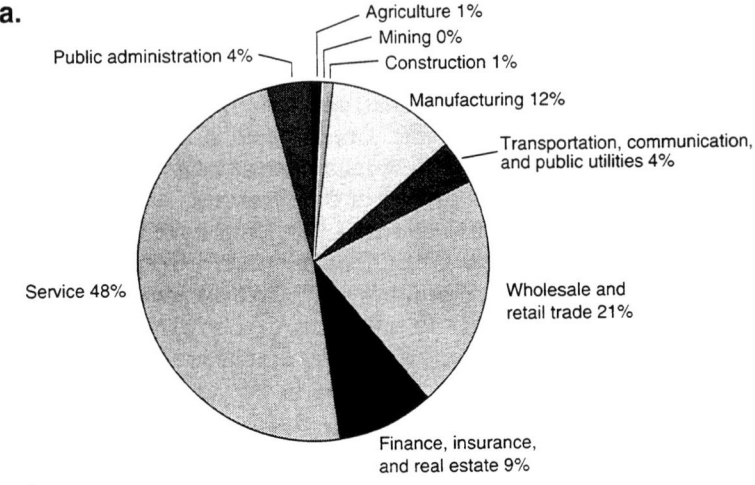

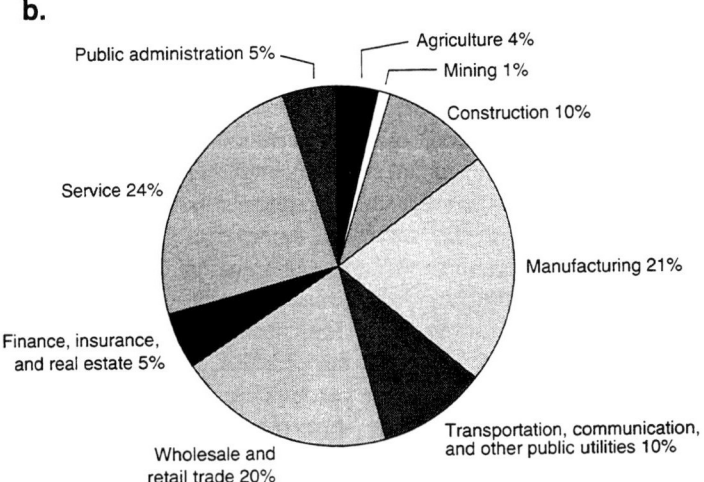

FIGURE B-1 Industrial employment of women and men by sector, 1992.

Historically, the complexities in characterizing "host factors" have led to a simple methodological solution: the wholesale exclusion of women from occupational health studies. A 1993 survey of the literature on cancer outcomes in occupational health found only a handful of studies in which women had not been excluded. In most cases their exclusion was based on a presumption of gender differences, on the idea that women are vulnerable or differentially susceptible and since their numbers were small, they should be excluded from studies. (Such presumptions have lead, as discussed above, to the idea that women should also be excluded from the actual employment situations themselves: that is, "protected" from the hazards.)

The actual state-of-the-art, however, is that, in fact, comparatively little is known about the actual exposures in the workplace. There are almost no registries with meaningful data on occupational exposures and illnesses; physicians are seldom trained to recognize environmental diseases. Those data which have been painstakingly collected in the past, such as the National Occupational Environmental Survey (NOES) and carried out by the National Institute for Occupational Safety and Health (NIOSH) are out of date. One priority, therefore, might be to provide adequate funding to re-establish a NOES-like survey and to establish a working conditions surveillance system under NIOSH's direction. Improvement of surveillance, one of the priorities in the new National Occupational Research Agenda (NORA), should also be integrated into the activities of other agencies that have an interest in biological responses to environmental exposures. In general, far too few resources have been devoted to the systematic collection and registration of occupational exposures and health outcomes in a way that is useful and conducive to research and establishment of meaningful policy.

Finally, a fundamental methodological issue confronting the panel is the "problem" statement itself: developing knowledge that will identify gender differences in susceptibility to environmental factors. The question of host-factor susceptibility is itself fraught with methodological issues with regard to determining when a problem is a woman's health problem versus when it is a public health problem that women also face. Asked another way, when are the observed differences attributable to gender (i.e., when are women differentially vulnerable?) and when does gender serve as a proxy variable for underlying exposures (i.e., when are women differentially exposed?)? Further, while there are real biological differences between men and women, it is also true that not all women or men are alike, nor is a woman the same biological creature at all stages in her life. Exposure to a given level of lead, for example, may have very different consequences for a healthy young woman than for an older worker with a 25-year lifetime exposure to lead.

Such variability in response greatly complicates the problem of how to extrapolate individual differences to differences in populations, particularly when the results of those extrapolations may influence the setting of standards. Standards are generally single values, by design, and as such they cannot capture the great variability in human response, unless they are set to accommodate the most sensitive individual. In some cases such standards would not be practical or feasible; in others such standards would unfairly exclude or discriminate against otherwise qualified individuals.

A variety of statistics illustrate the intervention of socioeconomic variables on women's health. For one thing, health is related to access to health care, which in turn is related to wealth: salary, insurance, pensions. Women are far less wealthy than men and have less access to health care, in general. For example, consider that approximately 80 percent of American women are employed outside the home, and up to 7 percent of these women work two jobs; but in no occupational category are women's average earnings more than 78 percent of those of men. Fewer women are covered by pensions or by group health insurance than men. Wealth alone, however, cannot explain every observed health difference, and the relationships between socioeconomic factors and health outcomes are complex as well. For example, higher levels of education are associated with a decreased incidence of cervical cancer and with a higher incidence of breast cancer but a lower breast cancer mortality rate.

Such complexity means that observed differences in health outcomes among women may be due to many factors other than "gender susceptibility," or they may be due to differential exposures. To a large extent these are separate issues. Even the term "exposure" is not clearly and consistently defined. When pursuing the topic of differential gender susceptibility, it is important to keep in mind the complexities of the conceptual framework and also the realities of the exposure situations at work. In the fourth edition of the International Labor Organization's *Encyclopaedia of Occupational Health and Safety*, several instances of gender differences in response are reported, but there is not a single instance of a working condition or exposure in which only females are adversely affected. From an historical perspective it is useful to recall the words of Anna Baetjer in her landmark 1946 study, *The Health and Efficiency of Women at Work*. She observed that the extent of illness among women workers was not known, largely because of a lack of data. She warned against broad generalizations about the weakness of women, when there were no underlying data to support such conclusions, a warning that is still appropriate today.

Environmental Exposure and Nutrition[2]

This topic involves many of the same paradigm issues developed in the preceding presentation: do women have different diets from men, as well as (or instead of) different environmental exposures, and does the interaction between diet and exposure increase or decrease their risks of disease? How does a woman's dietary intake influence her exposures to environmental factors? How do the patterns of exposure change over the lifespan, and how does a woman's changing physiology alter her exposure?

This topic does not exist as such in the nutritional literature, nor does it fit well in the priority-setting model of nutrition monitoring (see Figure B-2). This model is designed to identify current or potential public health problems based on levels of nutrient intake. From this perspective, the biggest threats to women are too much fat, cholesterol, and sodium (which increase their risk of cardiovascular disease and hypertension) and too little iron and calcium (which increase their risk of anemia and osteoporosis).

---

[2]This section is based on the presentation by Shiriki Kumanyika, Ph.D., professor of the Department of Human Nutrition and Dietetics, University of Illinois at Chicago.

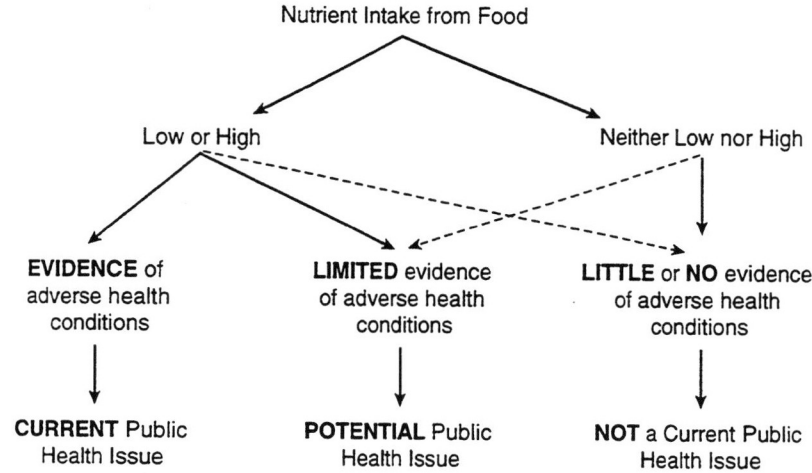

FIGURE B-2 Setting nutrition monitoring priorities.

Interagency cooperation is required to collect and link both kinds of data, but the result is the ability to generate data on dietary intake of contaminants. More effort is needed in this area.

How might researchers use the resulting data for surveillance: that is, how might women's dietary intake influence their exposures to environmental factors? One possibility could be that some foods that are eaten in greater quantities by women might serve as carriers of risk (toxicants or additives) or as carriers of protection (e.g., antioxidants). Or it may be that certain methods of food preparation predispose women to risk. Another possibility is that micronutrient deficiencies can predispose women to disease.

The differences in the recommended nutrient intakes for men and women are relatively small, with the exception of pregnant and lactating women (see Tables B-1 and B-2). Women do eat less food than men, however, which means they can have a harder time getting enough nutrients. Consequently, women's diets need to be better in order to be nutritionally adequate. Unfortunately, recent studies have shown that micronutrient intakes are below recommended levels in a sizable proportion of the population, including some nutrients with known or suspected protective roles against disease, such as vitamins A and E (cancers), vitamin B6 (heart disease), calcium (osteoporosis), and minerals (cardiovascular disease and immune function). Women are more likely than men to have low intakes of iron, calcium, and zinc. There is little information on whether low iron intake exacerbates problems.

TABLE B-1 Recommended Dietary Allowances

FOOD AND NUTRITION BOARD, NATIONAL ACADEMY OF SCIENCES–NATIONAL RESEARCH COUNCIL
RECOMMENDED DIETARY ALLOWANCES,[a] Revised 1989 (Abridged[1])

Designed for the maintenance of good nutrition of practically all healthy people in the United States

Category	Age (years) or Condition	Weight[b] (kg)	Weight[b] (lb)	Height[b] (cm)	Height[b] (in)	Protein (g)	Fat-Soluble Vitamins				Water-Soluble Vitamins							Minerals				
							Vita-min A (μg RE)[c]	Vita-min E (mg α-TE)[d]	Vita-min K (μg)		Vita-min C (mg)	Thia-min (mg)	Ribo-flavin (mg)	Niacin (mg NE)[e]	Vita-min B$_6$ (mg)	Fo-late (μg)	Vitamin B$_{12}$ (μg)	Iron (mg)	Zinc (mg)	Iodine (μg)	Sele-nium (μg)
Infant	0.0–0.5	6	13	60	24	13	375	3	5		30	0.3	0.4	5	0.3	25	0.3	6	5	40	10
	0.5–1.0	9	20	71	28	14	375	4	10		35	0.4	0.5	6	0.6	35	0.5	10	5	50	15
Children	1–3	13	29	90	35	16	400	6	15		40	0.7	0.8	9	1.0	50	0.7	10	10	70	20
	4–6	20	44	112	44	24	500	7	20		45	0.9	1.1	12	1.1	75	1.0	10	10	90	20
	7–10	28	62	132	52	28	700	7	30		45	1.0	1.2	13	1.4	100	1.4	10	10	120	30
Males	11–14	45	99	157	62	45	1,000	10	45		50	1.3	1.5	17	1.7	150	2.0	12	15	150	40
	15–18	66	145	176	69	59	1,000	10	65		60	1.5	1.8	20	2.0	200	2.0	12	15	150	50
	19–24	72	160	177	70	58	1,000	10	70		60	1.5	1.7	19	2.0	200	2.0	10	15	150	70
	25–50	79	174	176	70	63	1,000	10	80		60	1.5	1.7	19	2.0	200	2.0	10	15	150	70
	51+	77	170	173	68	63	1,000	10	80		60	1.2	1.4	15	2.0	200	2.0	10	15	150	70
Females	11–14	46	101	157	62	46	800	8	45		50	1.1	1.3	15	1.4	150	2.0	15	12	150	45
	15–18	55	120	163	64	44	800	8	55		60	1.1	1.3	15	1.5	180	2.0	15	12	150	50
	19–24	58	128	164	65	46	800	8	60		60	1.1	1.3	15	1.6	180	2.0	15	12	150	55
	25–50	63	138	163	64	50	800	8	65		60	1.1	1.3	15	1.6	180	2.0	15	12	150	55
	51+	65	143	160	63	50	800	8	65		60	1.0	1.2	13	1.6	180	2.0	10	12	150	55
Pregnant						60	800	10	65		70	1.5	1.6	17	2.2	400	2.2	30	15	175	65
Lactating	1st 6 months					65	1,300	12	65		95	1.6	1.8	20	2.1	280	2.6	15	19	200	75
	2nd 6 months					62	1,200	11	65		90	1.6	1.7	20	2.1	260	2.6	15	16	200	75

[1] NOTE: This table does not include nutrients for which Dietary Reference Intakes have recently been established [see *Dietary Reference Intakes for Calcium, Phosphorus, Magnesium, Vitamin D, and Fluoride, 1997*].

[a] The allowances, expressed as average daily intakes over time, are intended to provide for individual variations among most normal persons as they live in the United States under usual environmental stresses. Diets should be based on a variety of common foods in order to provide other nutrients for which human requirements have been less well defined. See text for detailed discussion of allowances and of nutrients not tabulated.

[b] Weights and heights of Reference Adults are actual medians for the U.S. population of the designated age, as reported by NHANES II. The median weights and heights of those under 19 years of age were taken from Hamill et al. (1979) (see pages 16–17). The use of these figures does not imply that the height-to-weight ratios are ideal.

[c] Retinol equivalents. 1 retinol equivalent = 1 μg retinol or 6 μg β-carotene. See text for calculation of vitamin A activity of diets as retinol equivalents.

[d] α-Tocopherol equivalents. 1 mg d-α tocopherol = 1 α-TE. See text for variation in allowances and calculation of vitamin E activity of the diet as α-tocopherol equivalents.

[e] 1 NE (niacin equivalent) is equal to 1 mg of niacin or 60 mg of dietary tryptophan.

TABLE B-2 Dietary Reference Intakes

FOOD AND NUTRITION BOARD, NATIONAL ACADEMY OF SCIENCES–INSTITUTE OF MEDICINE DIETARY REFERENCE INTAKES, 1997

Life-Stage Group	Calcium AI[a] (mg/day)	Phosphorus RDA[b] (mg/day)	Phosphorus AI (mg/day)	Magnesium RDA (mg/day)	Magnesium AI (mg/day)	Vitamin D AI[c,d] (μg/day)	Fluoride AI (mg/day)
Infants							
0 to 6 months	210		100		30	5	0.01
6 to 12 months	270		275		75	5	0.5
Children							
1 through 3 years	500	460		80		5	0.7
4 through 8 years	800	500		130		5	1
Males							
9 through 13 years	1,300	1,250		240		5	2
14 through 18 years	1,300	1,250		410		5	3
19 through 30 years	1,000	700		400		5	4
31 through 50 years	1,000	700		420		5	4
51 through 70 years	1,200	700		420		10	4
>70 years	1,200	700		420		15	4
Females							
9 through 13 years	1,300	1,250		240		5	2
14 through 18 years	1,300	1,250		360		5	3
19 through 30 years	1,000	700		310		5	3
31 through 50 years	1,000	700		320		5	3
51 through 70 years	1,200	700		320		10	3
>70 years	1,200	700		320		15	3
Pregnancy							
≤18 years	1,300	1,250		400		5	3
19 through 30 years	1,000	700		350		5	3
31 through 50 years	1,000	700		360		5	3
Lactation							
≤18 years	1,300	1,250		360		5	3
19 through 30 years	1,000	700		310		5	3
31 through 50 years	1,000	700		320		5	3

[a] AI = Adequate Intake. The observed average or experimentally set intake by a defined population or subgroup that appears to sustain a defined nutritional status, such as growth rate, normal circulating nutrient values, or other functional indicators of health. AI is utilized if sufficient scientific evidence is not available to derive an EAR. For healthy breastfed infants, AI is the mean intake. All other life-stage groups should be covered at the AI value. The AI is not equivalent to a RDA.
[b] RDA = Recommended Dietary Allowance. The intake that meets the nutrient need of almost all (97–98 percent) individuals in a group.
[c] As cholecalciferol. 1 μg cholecalciferol = 40 IU vitamin D.
[d] In the absence of adequate exposure to sunlight.

Significant threats such as environmental toxicants and carcinogens do not appear either in this model or on the list of public health issues related to nutrition, because of an underlying assumption that there are no toxicants and carcinogens in the food system. But while nutrition monitoring systems are not designed to measure contaminants, they do measure intake of the specific foods that might be contaminated, such as fruits and vegetables that might contain pesticides. EPA and FDA do monitor the food system for such contaminants; but the expectation is that if contaminants are discovered, they will be removed from the food system. Calcium intake is considered too low throughout the life cycle, a problem that is associated with increased risk of osteoporosis. Low calcium intake may also be associated with higher levels of lead uptake, but this connection has received very little study. Women are disproportionately affected by iodine deficiencies and goiter as well. Iodine deficiencies can lead to increased uptake of radioactive isotopes, which can be a problem in contaminated areas such as Chernobyl.

On the other hand, women (and especially older white women) are more likely than men to take vitamin supplements, which might mitigate these risks but might also introduce new risks. Women also appear to eat more fruits and vegetables than men—the so-called "salad factor"—although the data to support this perception are far from striking. Some of these foods may contain substances that protect against disease, such as antioxidants. However, the same fruits and vegetables may also contain contaminants, and this would expose women to higher risks. Conceptually, however, this type of increased susceptibility is behavioral rather than biological.

Some methods of food preparation, such as grilling or barbecuing, are suspected of inducing potentially dangerous enzymes and mutagens. However, it is difficult to identify major differences in food preparation between men and women. Women prepare most of the food, but in most cases they prepare it the same for themselves as for their families. Weight-consciousness has led men and women alike to avoid many fried foods.

Obesity may be more of a factor than undernutrition in susceptibility to environmental contaminants. Obesity has increased markedly since the 1970s, with one-third of American adults and one-fifth of adolescents now characterized as obese. Because obesity increases the amount of body fat, it provides greater reservoirs for lipophilic (fat-loving) toxicants that tend to reside in adipose tissue. Contrary to popular belief, however, by current definition (body mass index [calculated as weight in kilograms divided by the square of height in meters] of 25 or greater), being overweight is not a greater problem for women than men. Nevertheless, two of the differences between men and women that stand out in the national data are that women diet more often and consume more low-fat foods and beverages.

Women's nutritional status may have its own effect on their environmental exposures. Women are smaller than men on average, but have proportionately more adipose tissue and more cycles of fat gain and loss because of dieting behavior. To the extent that weight loss mobilizes potential toxicants that had been stored in adipose tissue, it may expose women to greater risk. This may explain why epidemiological data show an association between weight loss (voluntary or involuntary) and mortality.

The same may hold for other changes in body composition during pregnancy and lactation, menstruation, and menopause. Older women also experience greater bone loss than men, which may release metals and other substances that were stored in the bones. Fractures also release potential toxicants from the bones into the bloodstream, and this may expose women to greater risk if osteoporosis leads to more frequent fractures.

Review of the cancer literature reveals very little attention to these aspects of nutritional factors. Articles typically focus on the consumption of fruits and vegetables; to the extent that their antioxidant effects attenuate the risk of lung, cervical, renal cell, and nasopharyngeal cancers, dietary intake of these foods may reduce the risk of cancer. Many studies address environmental factors and many others consider diet, but very few consider both diet and environmental factors.

Multiple Environmental Exposures Across the Lifespan[3]

The first challenge in this area is to understand what is meant by "multiple exposures." Exposure to potentially harmful substances and stresses occurs not only in environmental and occupational settings but also in residential, recreational, and even medical settings. These separate exposures may have both cumulative and synergistic effects, and their combined consequences will be different at various points in a woman's menstrual cycle, career, and lifespan.

"Environmental" exposures, in the conventional sense, can include such sources as alcohol, water, air pollution, environmental tobacco smoke, and soil (especially for children). A few studies have examined various exposures in terms of differences between men and women. For example, male and female runners are impacted in different ways by the same levels of ozone, and women runners are impacted in different ways by ozone at different points in their menstrual cycle.

In the home, 90 percent of the work is still done by women. Consequently, in addition to their occupational exposures they also experience "residential" exposures. The agents for these exposures include cleaning products, some of which include solvents that can be absorbed through the skin. Other examples include pesticides and painting products. Men and women also have different hobbies; a recent study found that, among the top 10 hobbies for men and women, only one (bowling) was common to both. Hobbies such as gardening

---

[3]This section is based on the presentation by S. Katharine Hammond, Ph.D., associate professor of environmental health sciences at the University of California at Berkeley.

and furniture refinishing can lead to additional exposures, all of them independent of and additive to a woman's occupational exposures.

"Lifestyle" exposures can include such factors as smoking, drinking, and the use of recreational drugs. These exposures, in turn, can be modulated by the hormonal cycle. For example, one study showed that, given the same initial dose of cocaine, male subjects had over double the plasma level found in women. However, women in the follicular stage (before ovulation) have higher plasma levels of cocaine than those in the luteal phase, when estradiol and progesterone levels are higher (Lukas, et al., 1996).

Prescription and nonprescription drugs are used differently by men and women; they are also used differently by women at different times in their lives. This results in additional exposures, but it can also affect how the individual responds to other agents. For example, Tagamet (widely used by both men and women) also affects the p450 enzyme system, on which the body depends to detoxify some substances to which it is exposed. Nonsteroidal anti-inflammatory drugs are more commonly used by women, in whom arthritis is more common, and the effects of these drugs on the liver and kidney may interfere with their ability to detoxify some contaminants. Antidepressants are also used more often by women, but their interactions with other chemicals are unknown. More research is needed on patterns in prescribing drugs and on their interactions with other chemicals, as they relate to gender differences.

Hormonal interactions are different during prepuberty, puberty, adulthood, pregnancy, perimenopause, and menopause. Responses to chemical contaminants can vary accordingly. Artificial variations occur as a result of the use of birth control drugs and hormone replacement therapy. These variations are further complicated by temporal changes that are not captured by cross-sectional data. Between 1960 and 1993, for example, the proportion of women aged 25–29 who had no live births rose from 20 percent to 44 percent. At the same time, the onset of puberty is earlier and earlier. There is also a much higher level of smoking among teenage girls than there was a generation or two ago.

There have also been changes over time in the rates of occupational injuries: they are rising for women, declining for men (see Figure B-3). This is not the result of aging in women; instead, it is most likely due to the fact that, as women move into traditionally "male" occupations, they are also exposed to more risks. That is, their exposure has changed, not their susceptibility.

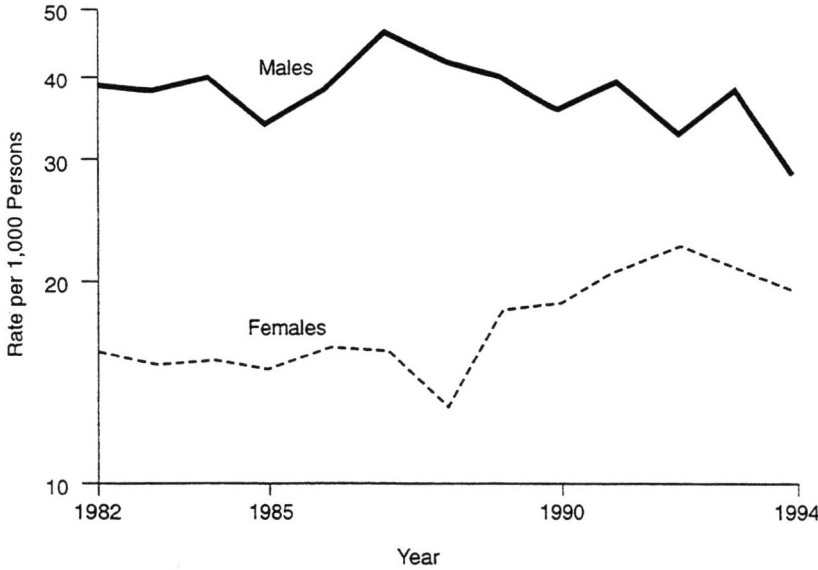

FIGURE B-3 Occupational injuries.

These multiple exposures are important because of the potential for synergy among them. A classic example is the relationship between exposures to smoking and asbestos and the risk of lung cancer (see Table B-3). People with neither exposure have a rate of lung cancer of 11 in 100,000. For asbestos workers the rate goes to 58 (a relative risk 5-fold greater than for those without exposure) and for smokers the rate is 123 (an 11-fold relative risk); but when people were exposed to both smoking and asbestos the rate is not additive (11 + 58 + 23 = 192), but it is over 600. Furthermore, the relative risk is not additive (5 + 11 = 16) but multiplicative (a relative risk of 5 × 11 = 55 times).

TABLE B-3 Synergistic Effect of Multiple Agents: Cigarette Smoking and Occupational Exposure to Asbestos

Lung Cancer Death Rate per 100,000 Man-Years	Nonsmokers	Smokers
Not exposed to:		
Asbestos	11	123
Asbestos workers	58	602

SOURCE: Hammond et al., 1979.

Exposures to "environmental factors," in this broader sense, differ not only between men and women but between home and workplace and over the lifespan. For example, women are more likely to be exposed to environmental tobacco smoke in the home, while men are more likely to be exposed in the workplace (see Figures B-4 and B-5). Similarly, the lifetime prevalence of psychiatric disorders is the same for men and women, but the distribution is different: substance abuse is higher in men, while anxiety and depression are higher in women. Presumably the drugs they are given are different as well. Men also respond to stress and competition by producing higher levels of epinephrine; this may also have significance in workplace challenges.

Because of these differences in the sources, levels, and combinations of exposures between men and women, estimates of the risks associated with chemical exposure cannot be based solely on adult male cohort studies. For example, current estimates of the risk of benzene exposure are based on such studies, but evidence from animal studies points to differential impacts on blood-forming tissues in male and female fetal, adult, and pregnant rats exposed to benzene and/or ethanol (see Table B-4). In adult males, all three exposure regimes suppressed the formation of erythroid colony-forming units (CFUs, the precursors of erythroblasts and erythrocytes). In females, on the other hand, no exposure had any effect on CFU formation in nonpregnant adults, but exposure to ethanol increased CFU formation in pregnant females, and the combination of ethanol and benzene increased CFU formation in females that had been exposed in utero. In this case, gender differences may actually be protective for women.

Similar evidence for the interaction of multiple exposures over the lifespan can be seen in a recent study of the link between smoking and breast cancer (Ambrosone et al., 1996). Previous epidemiological studies had been ambiguous; but this study focused on N-acetyltransferase-2 (NAT-2), an enzyme that is assumed to break down the aromatic amines in tobacco smoke, chemicals that are known to induce mammary cancers in animals and to cause DNA damage in human mammary cells.

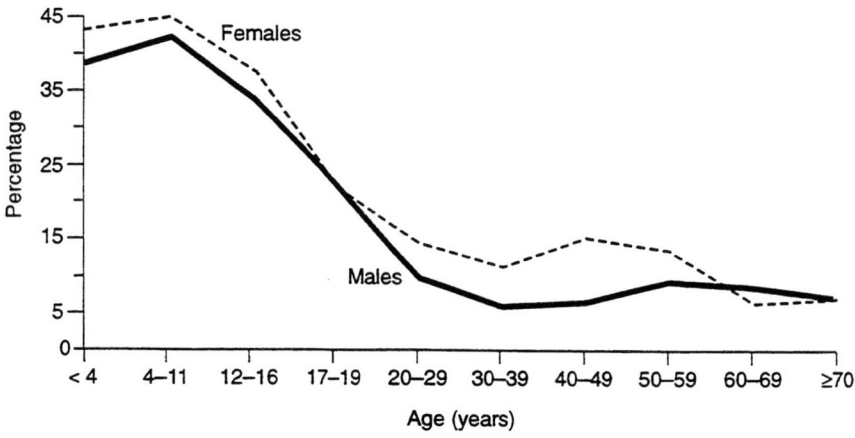

FIGURE B-4 Environmental tobacco smoke in the home.

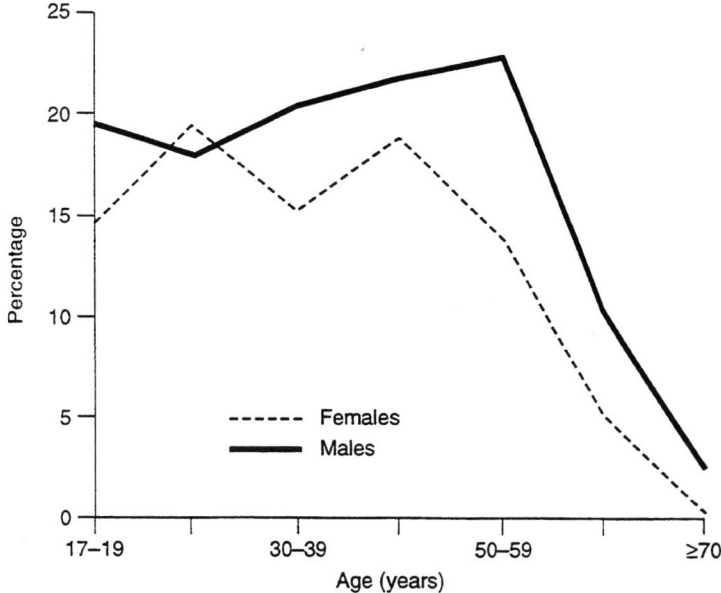

FIGURE B-5 Environmental tobacco smoke in the workplace.

TABLE B-4 Influences of Gender, Development, and Ethanol Consumption on Benzene's Effect on Erythroid Colony-Forming Units

Exposed Mice, 10-day Exposure	Benzene, 10 ppm	Ethanol, 5%	Benzene, Ethanol
Adult male	Reduced	Reduced	Reduced
Fetal male (days 6–15)	Reduced	—	—
Adult male exposed in utero	Reduced	Reduced	—
Adult female	—	—	—
Fetal female (days 6–15)	—	—	—
Adult female exposed in utero	—	—	Increased
Pregnant female (days 6–15)	—	Increased	Increased

SOURCE: Corti and Snyder, 1996.

Investigators learned that there are competing pathways for the metabolism of these compounds. The usual pathway is acetylation by NAT-2, which detoxifies them; individuals with low levels of NAT-2 are "slow acetylators" who clear these compounds more slowly and thus have higher residual levels. The competing pathway is oxidation by CYP-1A2, which can lead to a reactive

N-hydroxy metabolite that enters the circulation and is activated when it bonds to DNA in the target tissue. This would mean that subjects with low levels of NAT-2 would also have higher levels of this metabolite and thus be at even greater risk.

When investigators stratified their human subjects into those with high or low levels of NAT-2 (i.e., as rapid or slow acetylators), there was a much higher risk of cancer among postmenopausal women who were heavy smokers and slow acetylators (see Figure B-6). A slow acetylator who smokes a pack of cigarettes per day has an odds ratio for breast cancer that is 5.1 times that of a nonsmoker. Paradoxically, the odds ratio for a pack-a-day fast acetylator is less than for a nonsmoker, an apparent protective effect that deserves further study. The negative effects of smoking were also greatest among those who started smoking early, and for those who started before age 16 the odds ratio of slow to rapid acetylators was 4.5. The latter finding was consistent with the hypothesis that breast tissue is most sensitive to environmental factors during this time of development.

In a study of multiple exposures in the semiconductor industry, researchers followed 3,200 subjects in eight companies, equally divided between men and women and between fabrication and nonfabrication workers. Subjects were evaluated for 15 chemical agents, two physical agents, and numerous ergonomic stressors. Investigators found that fabrication workers (those in protective gear in the "clean-room" assembly areas) were more likely to be exposed to chemical and electromagnetic stressors than nonfabrication workers, regardless of gender. However, female fabrication workers were far more likely than males to be exposed to certain ergonomic stressors, such as the use of equipment with eyepieces (e.g., microscopes) and awkwardly placed vacuum wands. By using cluster analysis, researchers also found that fabrication workers were likely to be exposed to certain combinations of chemical agents. Given the possible interactions among these agents, cluster analysis techniques should be used more widely in epidemiological studies involving multiple exposures.

In conclusion, there are three dimensions in which exposure can change over the lifespan and which might warrant further research. First, there are variations in the chemicals to which we are exposed in a variety of settings. Second, there may be variations in our ability to absorb those chemicals. For example, gastrointestinal absorption of lead is twice as high in children as in adults. And third, there may be changes in our susceptibility to damage by these chemical agents. In utero, in particular, the DNA repair mechanisms, immune system, and blood-brain barrier are all poorly developed, leaving the fetus more vulnerable than an adult to many toxic and mutagenic compounds.

PANEL II: PATTERNS OF SUSCEPTIBILITY

The second panel addressed questions about the factors that evoke a response to exposure, the molecular and cellular processes involved in this response, and genetic differences in susceptibility to both exposure and response. Panelists also addressed the differences between men and women and among women in their responses to the same exposures. The first presentation focused on a disease that affects the human organism, while the others described laboratory studies that highlight the kind of basic research currently under way.

Epidemiology, Gender, and Environmental Influences on Multiple Sclerosis[4]

Multiple sclerosis (MS) is an autoimmune disease of the central nervous system that attacks the myelin sheath surrounding the spinal cord. There are 350,000 cases in the United States, making it (after trauma) the second most common neurological cause of disability in young adults. The typical patient is an otherwise healthy, young white woman, although the disease also occurs in men and less frequently in all racial groups. The incidence of MS appears to be rising, especially among women. There are new techniques for diagnosing the disease, but there is no specific laboratory test for MS. As a result, there may be a large number of patients who have MS and are never diagnosed properly. Autopsies suggest that perhaps 25 percent of all cases are "silent."

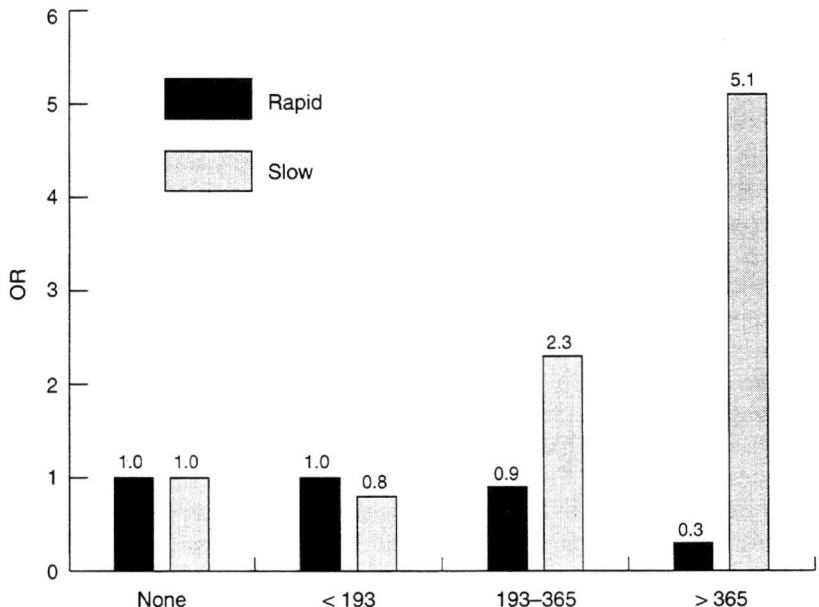

FIGURE B-6 Smoking and the risk of cancer among postmenopausal women.

---

[4]This section is based on the presentation by Peter Riskind, M.D., chief of neuroimmunology at Massachusetts General Hospital.

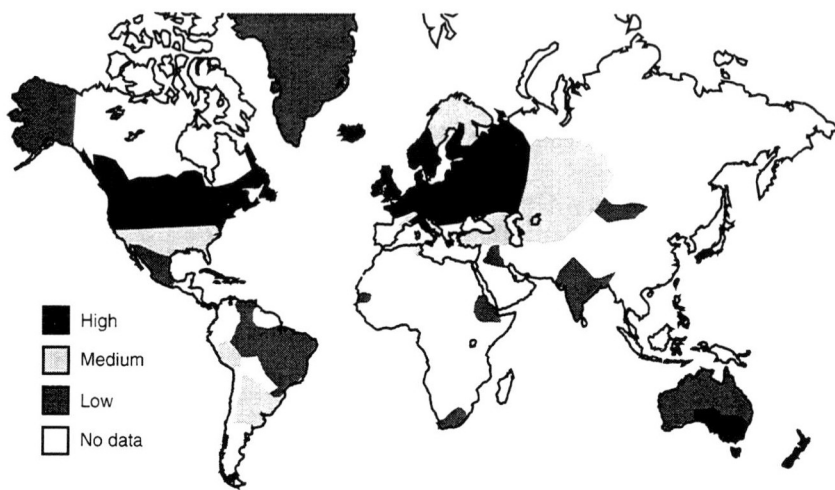

FIGURE B-7 Geographic localization of multiple sclerosis in the world.

Data suggest that the risk of developing MS is higher in white women as compared to white men by a factor of almost two; but white men have a higher risk than black men, and black women have a lower risk than white women (Kurtzke, et al., 1979; Kurtzke, 1977). In general, whites have a higher risk than blacks. Worldwide, there is a very low rate of multiple sclerosis in Africa and in Asia (See Figure B-7). There is a distinct localization of MS in the temperate latitudes: Europe (especially Scandinavia), North America, and Australia. In the United States, the highest incidence is in the northern tier of states, possibly because of the high concentration of Scandinavian immigrants in the northern states, plus some unknown environmental factor. Other known risk factors include average temperature, dairy products, meat, and sanitation. Common characteristics associated with higher risk of MS are urban residence, high socioeconomic status (SES), Swedish ancestry, and high education attainment (Kurtzke and Page, 1997). A man with all these risk factors might have up to 119 times the risk of contracting MS as an average American would; a woman with these risk factors, perhaps 200 times the risk.

Evidence from the Faroe Islands suggests that some infectious agent is involved in the etiology of MS (Kurtzke, et al., 1995), and several viral diseases demonstrate similar geographical patterns. Multiple sclerosis is a disease of humans; however, for research purposes, the animal model for MS is experimental autoimmune encephalomyelitis (EAE). Transgenic mice that develop spontaneous EAE are less likely to develop the disease in a germ- and virus-free environment, indicating that an infectious agent may be a cofactor.

Evidence for some genetic contribution to MS comes from its greater incidence among Caucasians as opposed to African Americans; its association with certain histocompatibility subtypes (e.g., HLA subtype DR2); and the way it is manifested among twins: 25 percent of identical twins also develop the disease, versus 4 percent of fraternal twins. Although these studies indicate some genetic contribution, the major susceptibility factor seems to be environmental.

Like most autoimmune diseases, MS is more common among women than men: approximately two-thirds to three-quarters of all MS patients are women. The age of onset is also earlier among women than men. However, among the men who do contract MS, the disease is more severe than among women, and the 10-year mortality rate is higher.

Sex hormones appear to have an important effect in MS. Pregnancy has a major effect, with a marked decrease in the number of relapses in the third trimester, at a time when female sex hormones are very high. Estrogen may be a protective factor. There may also be a role for other hormones, such as pituitary and adrenal gland hormones, that are different between men and women. In the animal model, for example, the suppression of prolactin (a pituitary hormone which is secreted in higher levels in females) alleviates the severity of EAE.

The effect of gender on susceptibility is confounded by environmental factors in complex ways, in both humans and animal models, and gender differences often disappear with changes in environment. Like breast cancer, MS is more prevalent in high-SES women; while lifestyle factors are difficult to isolate, there does seem to be an association between MS and a diet high in dairy and meat.

Estrogen Receptor Knockout Mouse Studies and Implications for Differences in Susceptibility[5]

Researchers at the National Institute of Environmental Health Sciences (NIEHS) have focused on the basic mechanism of action of estrogen, in order to understand where environmental estrogen might disrupt normal physiological processes. Estrogen produces a wide range of responses in a variety of sites, including the cardiovascular system and bones, in the male as well as the female. A variety of compounds in the environment can have estrogen-like effects on the body.

---

[5] This section is based on the presentation made by Kenneth Korach, Ph.D., Scientific Program Director of the Environmental Diseases and Medicine Program at the National Institute of Environmental Health Sciences.

Endocrinologists believe that the intracellular action of estrogen—whether synthetic, endogenous, or environmental—is a receptor-mediated pharmacological reaction. At the molecular level, estrogen may also operate through non-receptor-mediated action, with productive or antagonistic effects relative to the genomic action. By developing an estrogen-receptor knockout (ERKO) mouse model, researchers have been able to differentiate and study the receptor-mediated actions of estrogen. A second estrogen receptor, ER-beta, has recently been discovered; ER-beta is still present in the ERKO mouse but its independent role has yet to be studied.

When treated with three different types of estrogen agonists (e.g., stilbene, steroidal, and triphenylethylene), the classic uterotropic bioassay shows that ERKO mice are unresponsive to the hormone treatment. This also demonstrates that the uterine hyperemia measured by this assay is a receptor-mediated response. The receptor itself is a ligand-activated transcription factor. Another very sensitive marker of response is the stimulation of lactipherin, whose gene response and mRNA increase about 300-fold in response to a single injection of hormone; but ERKO mice are totally unresponsive, again demonstrating the total lack of functional estrogen receptors.

The ovary is dramatically affected by the absence of estrogen receptors. In the ERKO animals, follicular genesis does not proceed past the secondary follicular stage, and the ERKO ovaries never ovulate. As a result, the ERKO animals are infertile. Current studies indicate that the granulosa cells in the follicles undergo increased apoptosis, causing the follicle to degenerate and be resorbed prior to maturation. This may provide an animal model of polycystic ovary syndrome, a clinically interesting possibility that researchers will evaluate further.

Estrogen is also involved in the expression of the progesterone receptor, which is implicated in mammary tumor and breast cancer studies. However, the ERKO mouse shows no stimulation of the progesterone receptor, which at least in the ovary appears to be totally dependent on a functional estrogen receptor. In the uterus, on the other hand, there is both estrogen-dependent and estrogen-independent expression of the progesterone receptor. For some reason, the regulation of this receptor is different in these two reproductive tract tissues.

Hormone levels are dramatically altered in ERKO females. They have extremely high circulating levels of estradiol, because they lack negative feedback systems. Castration ablation experiments in males show that their gonadotropins go up as well, but in intact ERKO males they remain in the normal range for wild-type animals. This suggests that there may be a difference in regulation of gonadotropin between males and females, as well as a difference in the specific gonadotropin secretion: that is, serum LH levels are elevated but serum FSH is not.

Because the ERKO female does not develop mammary gland tissue, this model allows us to examine the role of the estrogen receptor in the development of breast cancer. Researchers have done this by crossing the ERKO mouse with the WNT-1 mouse, which has a high susceptibility to mammary cancer. The results show that the rudimentary duct tissue of the ERKO mouse is still susceptible to the action of the oncogene. That is, the WNT oncogene does not require a functional estrogen receptor to produce its phenotype. A surprising result was that ERKO females had a 58 percent incidence of tumors, while wild-type males had a 49 percent incidence, which may indicate that, in this model and with this oncogene, the female has increased susceptibility or the male has some protective effect. Further testing with other oncogene crosses is currently in progress to evaluate this finding.

In the ERKO male, there is extensive dismorphogenesis and swelling of the testes and a lack of sperm cells over time. Sperm count and sperm motility both decrease, and the remaining sperm are incapable of in vitro fertilization. In addition, the ERKO male's brain has tyrosine hydroxylase levels comparable to a wild-type female, indicating that the brain has also been reprogrammed. ERKO males have also lost their aggressive behavior; and the reason may be that male androgens, chemically transformed to estrogen, are responsible for male behaviors. The loss of these behaviors in ERKO males suggests that chemical transformation requires a functional estrogen receptor to produce male behavioral phenotypes.

Researchers have just begun to examine the effect of environmental estrogens. Because both male and female ERKO mice are infertile, researchers must generate the recessives from inbreeding of heterozygotes. This takes a lot of time, and researchers are only now getting a large enough pool of animals to do further treatment studies. Preliminary data indicates that Genistein may produce growth effects through the estrogen receptor in the uterus, but the regulation of LH negative feedback may involve a nonestrogen receptor mechanism.

Gender Differences in Metabolism and Susceptibility to Environmental Exposures[6]

Recent data indicate that the relatively small gender-specific differences in the metabolism of xenobiotics do not fully explain the gender-specific adverse effects of toxicants. Nor does the current literature indicate that gender-specific differences in the induction of catabolic enzymes by toxicants are responsible for the gender-different toxic effects that are observed. More importantly, gender differences are observed in isolated cells which possess little or no capacity to metabolize xenobiotics. This is not to say that gender differences in metabolism or enzyme induction do not exist, but rather to demonstrate the basis for a hypothesis that there are gender-specific differences in sensitivity and in the mechanism of toxic action, which are related to the primary rather than a secondary effect of a compound. On this basis, a hypothesis has been tested, using a well-defined toxicant, the herbicide 2,3,7,8-tetrachlorodibenzo-p-dioxin (TCDD, a known

---

[6]This section is based on the presentation made by Bill L. Lasley, Ph.D., professor of environmental health and reproduction at the University of California at Davis.

carcinogen and hormone disruptor in rats). TCDD's induction of adverse effects in adipose cells, a gender nonspecific tissue, was then studied. Metabolism is not the primary issue in these studies, since TCDD, with a half-life of several years in humans, never clears the system during these studies in either gender, and studies show no pharmacokinetic differences between young and old animals.

The following data are consistent in rat, hamster, and monkey models; they support the concept of a gender-specific difference in the sensitivity to some xenobiotics as well as a gender-specific difference in the mechanism of toxicity. These data also suggest that adipose tissue should be added to the list of sex-steroid-hormone target tissues. These data also predict that some toxicants should have a greater effect on lower vertebrates, which are more sensitive to sex steroids in terms of somatic development.

The underlying hypothesis of our recent research is that some signal transduction receptors and pathways evolved prior to sexual reproduction: that is, they are "pregender." These ancient transduction pathways have been re-adapted to other functions after sexual reproduction arose. This concept fits with the observations that hormones do not change as much in evolution as does the use to which they are adapted. Such re-adaptation may have paralleled and played a role in the expression of receptors and transduction pathways which became central for sex-steroid-hormone transduction and reproductive processes. We speculate and attempt to provide evidence that some toxicant-sensitive transduction pathway, overlap with sex-steroid-hormone-induced transduction pathways. Our data support the concept that some toxicants interrupt the mechanism by which sex-steroid-hormones program cells to be gender-differentiated and to function in predicted gender-specific ways. Some of our experimental data indicate that toxicant transduction pathways overlap with steroid and growth factor pathways, and this overlap may represent the basis of gender-specific differences in susceptibility to toxicants.

Pivotal to our general hypothesis is the observation that some currently accepted transduction pathways for toxicants evolved prior to expression of sexual reproduction. The eight-cell mouse embryo demonstrates that the AH receptor, which is a receptor for TCDD-like toxicants, exists prior to the estrogen receptor (Peters and Wiley, 1995). This observation is consistent with the concept that the AH receptor evolved prior to the estrogen receptor and prior to sexual reproduction. This lays the foundation for understanding gender differences, which is expanded below.

When intact monkey granulosis cells are exposed in vitro, TCDD decreases the level of MAP-2 kinase (which transduces growth factor pathways). In contrast, exposure to estrogen increases MAP-2K, and TCDD blocks this action of estrogen. When the nucleus is removed from the cells, the effects of TCDD and estrogen on the nucleus-free cell system are the same. Tamoxifen (an antiestrogen that acts at the level of the estrogen receptor) blocks the effect of estrogen but not the effect of TCDD, suggesting that TCDD is not operating through the estrogen receptor but through the AH receptor. These data also imply the existence of prenuclear (cytosolic) effects of both TCDD and estrogen which may be estrogen-receptor independent, and that cytosolic pathways may be involved in some gender-specific effects.

In whole-animal studies, TCDD decreases the growth rate of immature female rats but not the growth rate of mature females or castrated males treated with estrogen, suggesting that mature females are protected from the negative effects of TCDD by estrogen. These data alone do not prove that sex steroids such as estrogen determine gender-specific toxicities; but this model makes it possible to address questions about the mechanistic basis for (1) gender differences in sensitivity to environmental factors, (2) the increased sensitivity of dams to some toxicants during pregnancy, (3) steroid hormone disruption by environmental factors, and, if one extends the concept, (4) the underlying basis for differences in species: that is, sensitivity in the induction of developmental defects. There is little hard evidence to support the concept that overlapping pathways are the basis for the observed developmental defects, but it is often overlooked that the development of fish, reptiles, and birds are more susceptible to environmental and steroid inducers than are mammals.

Existing data make it possible to test the following specific hypothesis: *sex steroid hormones and growth factor transduction pathways are shared by toxicant-induced pathways.* Sex-steroid (and possibly growth factor) modulation of these pathways determines the effect of the toxicant. These pathways can originate in the cytoplasm and involve phosphorylation/phosphatase activities.

Our current data demonstrate that many of the observed gender differences in response to toxicants are qualitative (i.e., mechanistic) rather than quantitative (sensitivity alone) differences. These differences may have been previously overlooked because males are often used in mechanistic studies to avoid the confounding affect of cyclic steroid hormone levels in females. In addition, not all environmental toxicants in the same class of compounds (e.g., pesticides including arylhydrocarbons, chlorinated hydrocarbons) will have different gender-specific adverse effects, because receptor-ligand interactions may be as structure-specific for toxicants as they are for sex steroids. While both gender-specific steroids and toxicants ultimately exert a portion of their effects through the interaction at the nuclear level, many of these actions are initiated and dependent on cytosolic events, but some are nuclear independent. While some of these cytosolic actions depend on cytosolic receptors, it is possible that some do not. Evidence to support the prenuclear nature of these pathways comes from studies of the effect of estrogen and TCDD on three cytosolic enzymes: tyrosine kinase, MAP-2 kinase, and PKA. It is clear that nuclear pathways are affected downstream of these cytosolic pathways, as shown by studies of the effect of estrogen and TCDD on levels of AP-1, and underscore how toxicants may disrupt steroid hormone and growth factor transduction pathways.

These and other observations lead to a second hypothesis: *estrogen has both immediate and long-term effects on cytosolic signal transduction pathways.* Depending on cell type, estrogen will enhance or dampen the adverse effect of the toxicants that employ these pathways. The positive and negative effects of estrogen are time- and dose-related and appear to require the presence of the estrogen receptor. The latter point is worth stressing: low levels of estrogen, unopposed for a prolonged time period, may have the same effect as a higher dose for a short interval. This effect is demonstrated in studies of castrated female rats, in which exogenous estrogen replacement (simulating the long-term effect of low levels of estrogen seen with maturation) protects the animals against the adverse effect of TCDD (e.g., the loss of body weight; see Figure B-8) which is seen in males and untreated females. These data also provide evidence that transduction pathways, altered by estrogen in a relatively short period of time, interact with the mechanisms associated with the adverse effects of the toxicant.

Androgens also appear to modulate the effects of toxicants, either directly (through the androgen receptor) or by antagonizing the action of estrogen. As evidence of this, TCDD causes an increase in tyrosine kinase activity in the intact mature male rats, but this effect disappears in castrated males, which look strikingly similarly to intact females (see Figure B-9). This suggests that testosterone enhances some effects of TCDD, while estrogen protects or attenuates them. Similar results have been obtained in guinea pigs and suggest that the modulatory effects of androgens may be as important, if not more important, as estrogens; however, genomic differences cannot be ignored and caution should be exercised in interpreting data from experiments in which either gender is treated with sex-steroid hormones to elicit a specific response.

Taken together, these results permit the posing of a third hypothesis: *one type of hormone disruption occurs when toxicants and hormones (and perhaps growth hormones as well) share a critical transduction pathway and have opposing effects on that pathway*. Evidence to support this hypothesis comes from studies of granulosa cells, in which polyaromatic hydrocarbons—TCDD in this case—decrease the cell's ability to produce estradiol and also reduce estrogen receptor levels, thus demonstrating two different "hormone disrupting" mechanisms. Consequently, TCDD-like compounds can disrupt both estrogen production and estrogen action in the same or different cells. If the presence of estrogens and the ability for them to act are protective in terms of some adverse effects, then we can expect, and will find, delayed effects of some toxicants in females which are similar to the acute effects seen in males. These results suggest that TCDD-like compounds interfere not only with the transduction pathway but also with signaling pathways and have an important time-dependency in terms of adverse effects.

Finally, with regard to the adverse effects of environmental toxicants on development, data thus far point to a fourth specific hypothesis: *lower vertebrates are more sensitive to the adverse effects of toxicants on development, because they remain more susceptible to steroid hormone-induced developmental change*. Some lower vertebrates may be more hormonally programmed and less genetically programmed to develop normally, compared to mammals. For example, the sex ratio in alligators depends on the temperature at which the eggs are incubated, and when bird eggs are given testosterone the entire clutch develops as males. In mammals, on the other hand, if the SRY gene is present, the individual will be programmed to differentiate as a male. A consequence of

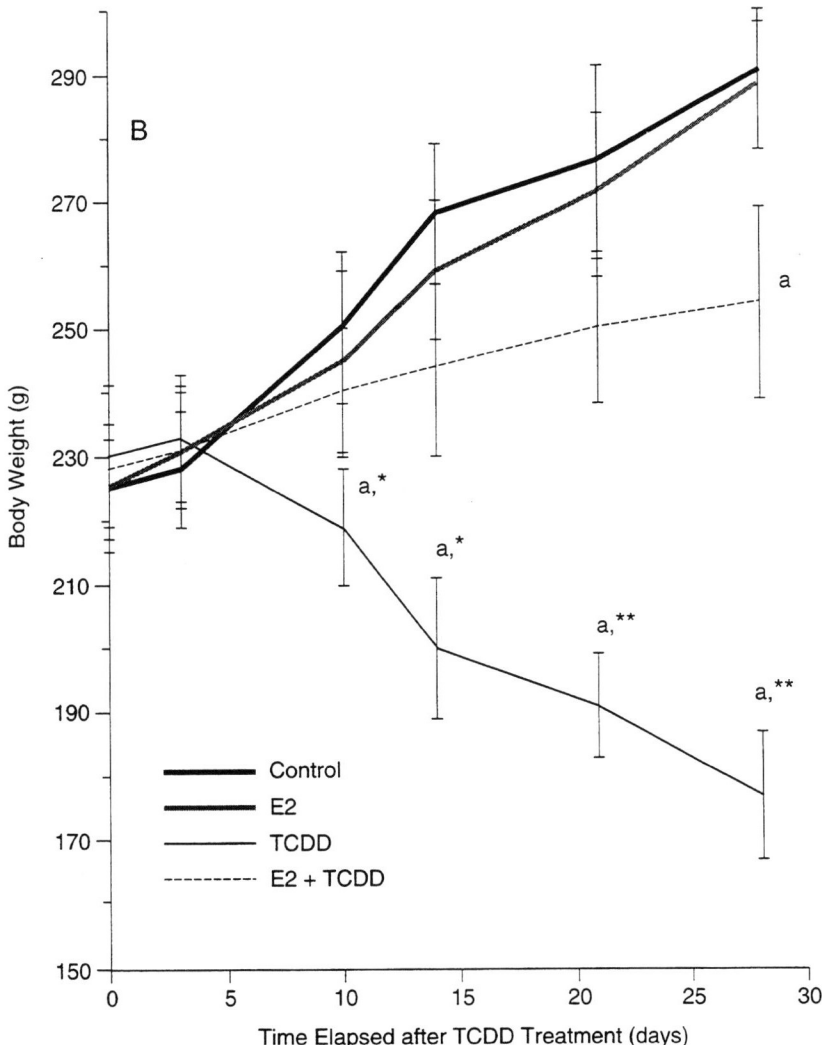

FIGURE B-8 Estrogen protective effects against TCDD.

this is that levels of toxicants that produce major developmental changes in lower vertebrates have less effect on the reproductive development of rats. This concept may help explain why wildlife biologists see environmental estrogens as a much greater threat than do experimental toxicologists. The mechanisms for these adverse effects, however, may be similar in lower and higher vertebrates, with the primary difference being the degree of genetic programming for development and differentiation. We hypothesize that the major pathway for these interactions appears to be MAP-2 kinase, a very early cytosolic pathway that overlaps with pathways that, later in evolution, are used for sexual development and differentiation. This explanation relies on the previously speculated interactions between TCDD-like compounds and estrogen and also on interactions at the level of helper proteins in the nucleus and factors that control cell cycles. These same concepts can be extrapolated from developmental defects to interactions between and among steroid hormones, growth factors, and toxicants and to changes in the proliferative potential of certain cell types and the induction of precancerous states. Cells may be likely to be more (or differentially) sensitive to toxicants early in their development and become less sensitive to toxicants as they reach their end-point differentiation. Comparative studies may be useful for examining these issues, as somatic cells from lower vertebrates have more plasticity in terms of responding to steroid induction than do somatic cells from higher vertebrates.

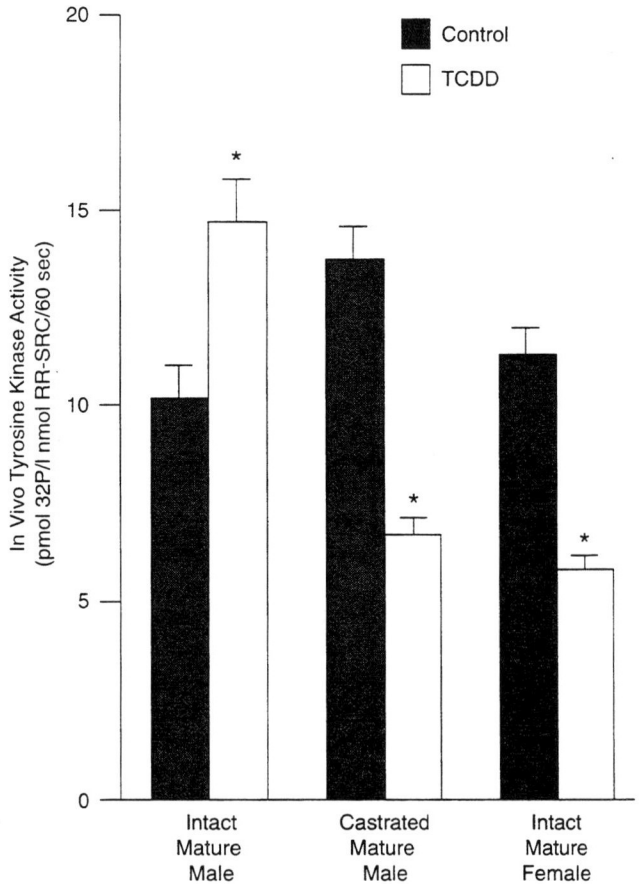

FIGURE B-9 TCDD and tyrosine kinase activity.

Molecular Markers of Carcinogenesis: Gender Differences[7]

Traditionally, environmental epidemiology has relied entirely on measurement of external exposure. Molecular epidemiology, on the other hand, has allowed a fuller understanding of the complex biochemical and genetic changes that occur as a result of exposure, up to and including clinical disease (see Figure B-10). The goal is to provide prevention and intervention measures that will reduce the likelihood of disease—in this case, cancer. These techniques also raise a number of socioethical concerns, such as discrimination in employment or health insurance, that must also be addressed.

These techniques include not only markers to measure the internal dose but also markers of actual biological effects, notably changes in DNA structure. This follows the paradigm that genotoxicity is the hallmark or driving mechanism of cancer. Assays developed in cell or animal studies have now been applied to humans to provide markers of early biological effects that will lead to cancer.

Researchers are also focusing on the role of oncogenes and tumor suppressor genes as markers of susceptibility. These are being studied not only for their role in carcinogenesis but also for racial, ethnic, and gender differences. Research should also address the effect of endogenous agents (e.g., steroidal hormones) in modulating the activities of these genes.

The availability of these genetic markers complicates the design of epidemiological studies by adding new levels of complexity with regard to exposure, disease, and genetic susceptibility. Even this level of variables is rather simplistic, given the polymorphism of human genes, but these considerations have not always been considered in ongoing studies. Sibling-paired twin studies could narrow the regions and chromosomes on which we look for a polymorphic gene that may or may not be responsible for disease. The pursuit of the BRCA-1 gene is a classic example of this approach. Once the population impact of gene expression is discovered (or determined) by epidemiologic studies, however, it falls to the laboratory scientist to look for functional mutations and to discover the mechanism by which the gene leads to disease.

Animal studies have demonstrated gender differences in susceptibility to cancer in specific nongonadal tissues and in response to specific categories of carcinogens. Genetic factors also appear to play a role, as in the ERKO mouse described earlier. Mechanism studies suggest that both hormones and receptors play a role in these differences, and that exogenous hormonal mimics can modulate both endocrine and metabolic pathways.

---

[7]This section is based on the presentation by Greg Cosma, Ph.D. assistant professor of toxicology at Colorado State University.

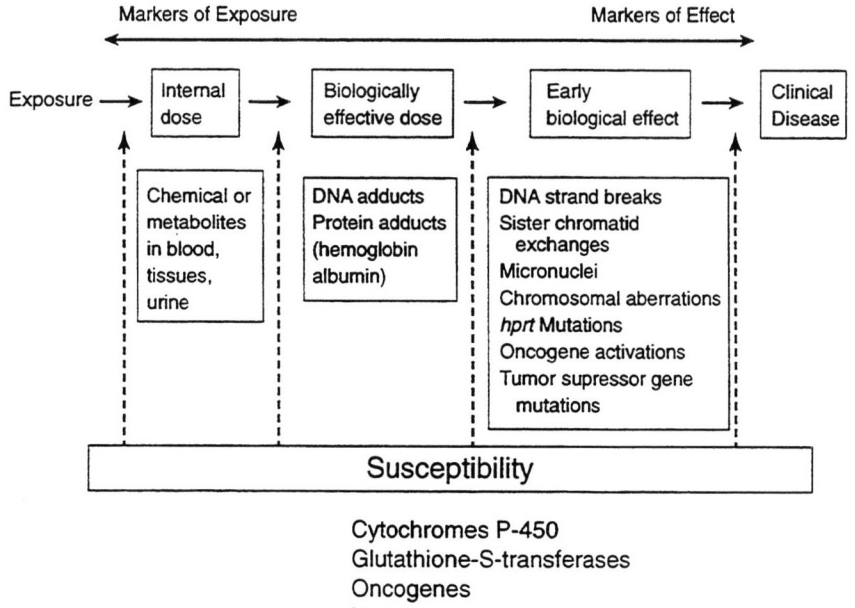

FIGURE B-10 Changes that occur as a result of exposure.

Human studies have also shown gender differences in susceptibility to some environmental exposures, in terms of both odds ratio and target organ (see Table B-5). The tobacco data in particular are disturbing, given the recent increase in tobacco usage by adolescent females. The role of pesticides and other xenoestrogens is also being investigated in several laboratories. The epidemiologic studies that showed association between organochloride pesticides and breast cancer were necessarily limited by their retrospective study designs. Estimates of individual exposure had to be reconstructed from historical or imputed data and all possible risk factors for breast cancer could not be accounted for. Therefore, these studies need to be followed up by prospective epidemiology studies as well as laboratory investigations of the associations.

Studies of molecular biomarkers of susceptibility have found some gender differences in baseline cytogenetic markers and DNA adducts, particularly with regard to tobacco and lung cancer. Researchers found race- but not gender-related differences in the frequency of polymorphisms in genes related to Phase I metabolism, which activates procarcinogens into carcinogens. However, females do have a higher incidence of polymorphisms in certain genes related to Phase II metabolism, which has a role in detoxifying or eliminating carcinogens, and in the p53 tumor suppressor gene. Collectively, these studies show clear gender differences in both susceptibility and the frequency of markers of susceptibility, with racial and ethnic differences among women with regard to markers. In particular, they also suggest that women have a higher susceptibility than men to lung cancer following exposure to tobacco.

DES has been used as a model of environmental estrogen, and there are reports of reproductive abnormalities in the male offspring of DES subjects, but there has been little tissue-specific study of the effects on the male. It may be that there are "windows of susceptibility" during development, and new studies will address susceptibility at different stages of pregnancy. In general, males appear to be less sensitive to estrogenic compounds, at least initially, but they have a very steep dose-response curve and an abrupt response at higher levels. This may be because androgen is protective in the male, whereas estrogen exaggerates adverse effects.

This review suggests five areas where further research is needed:

1. inclusion of women in occupational studies and further identification of environmental risk factors;
2. further clarification of gender differences in frequency of known markers of genetic susceptibility;
3. evaluation of steroid receptor variants and susceptibility to environmental cancers;
4. vigorous application of animal models to study underlying regulation of environmental carcinogenesis; and
5. identification of biological causality between genetic susceptibility markers and gender-related human cancers.

TABLE B-5 Gender Differences in Cancer Susceptibility: Human Studies/Environmental Exposures

- Tobacco-related cancers (ORs) (Zang and Wynder, 1996)
  Lung (bronchogenic carcinoma)—female: 8.1, male: 4.6
  Oral—female: 5.0, male: 2.0
- Dioxin-related cancers (Seveso, Italy) (Landi et al., 1997).
  — Men: Leukemia, esophageal, rectal
  — Women: Liver, stomach, colon (decrease in breast cancer)
  — Hodgkin's disease: Women > men
- Pesticide exposure (Zahm et al., 1994)
  — Non-Hodgkin's lymphoma: Women < men
  — Soft-tissue sarcomas: Women < men
- Pesticide-breast cancer controversy

C

Acronyms, Abbreviations, and Glossary

ACROYNMS AND ABBREVIATIONS

AH	aryl-hydrocarbon hydroxylase receptor
CDC	Centers for Disease Control and Prevention
DNA	deoxyribonucleic acid
DOD	Department of Defense
EAE	experimental autoimmune encephalomyelitis
EPA	Environmental Protection Agency
ER	estrogen RNA
ERKO	estrogen-receptor knockout
HHS	Department of Health and Human Services
IOM	Institute of Medicine
HLA	histocompatibility
LH	luteinizing hormone
mRNA	messenger RNA
MS	multiple sclerosis
NCHS	National Center for Health Statistics
NCRR	National Center for Research Resources
NHANES	National Health and Nutrition Examination Survey
NIEHS	National Institute of Environmental Health Sciences
NIH	National Institutes of Health
NIOSH	National Institute for Occupational Safety and Health
NRC	National Research Council
ORWH	Office for Research on Women's Health, NIH
RNA	ribonucleic acid
SES	socioeconomic status
TCDD	herbicide 2,3,7,8-tetrachlorodibenzo-p-dioxin
VA	Department of Veterans Affairs
WHI	Women's Health Initiative

GLOSSARY

Environment is comprised of all chemical, physical, and biological features of the earth than can affect or be affected by human activities.

Environmental exposure occurs in a variety of ways: in different settings (e.g., the home, the workplace), through different routes (e.g., foods), because of different activities (e.g., chores, hobbies), or because of unique or critical times in the lifespan.

Gender is used when referring to the social expression of living with one or two X chromosomes.

Gender differences are primarily determined by non-biologic factors, such as social roles, but influenced by sex-steroid hormone metabolism, anatomy, immunologic function, and genetic influences.

Genes are the fundamental physical and functional unit of heredity. A gene is an ordered sequence of nucleotides located in a particular chromosome that encodes a specific functional product.

"knockout" mice are experimental mice created by disrupting (knocking out) the function of a specific gene.

MAP-2 kinase is an enzyme that transduces growth factor pathways.

Multiple sclerosis is an autoimmune disease of the central nervous system that attacks the myelin sheath surrounding nerve fibers in the brain and the spinal cord.

Mutation is a permanent, transmissible change in the DNA sequence. It can be an insertion or deletion of genetic material or an alteration in the original information.

Polymorphisms are naturally occurring variations in a DNA sequence. Polymorphisms are useful markers because they allow researchers to distinguish between DNA of different origins.

Sex is generally used to designate the chromosomal or biologic phenomena linked to having one or two X chromosomes. Normal females have two X chromosomes, while normal males have one X chromosome and one Y chromosome.

Susceptibility is the state of being readily affected or acted upon by the environment. The impact depends on exposure and the individual's ability to respond.

TCDD is the herbicide 2,3,7,8-tetrachlorodibenzo-p-dioxin, a known carcinogen and hormone disrupter in rats.

Transgenic mice are mice that have a foreign gene introduced into their cells.

D

Workshop Agenda, Speakers, and Participants

Gender Differences in Susceptibility to Environmental Factors
May 21, 1997
Washington, D.C.

8:00 a.m.   Welcoming Remarks and Introduction
                   Nancy Fugate Woods, University of Washington, Chair of Committee
                   Valerie P. Setlow, Director, Division of Health Sciences Policy

Panel I: Patterns of Exposure Among Women

8:30   Environmental Exposure in the Workplace
            Jeanne M. Stellman, Columbia University

9:00   Environmental Exposure and Nutrition
            Shiriki Kumanyika, University of Illinois at Chicago

9:30   Multiple Environmental Exposures and the Lifespan
            S. Katharine Hammond, University of California at Berkeley

10:00   Response and Discussion
            Eula Bingham, University of Cincinnati
            David H. Wegman, University of Massachusetts

10:30   Break

Panel II: Patterns of Susceptibility

10:45	Epidemiology, Gender, and Environmental Influences on Multiple Sclerosis *Peter N. Riskind, Massachusetts General Hospital*
11:15	Estrogen Receptor Knockout Mouse Studies and Implications for Differences in Susceptibility *Kenneth S. Korach, National Institute of Environmental Health Sciences*
11:45	Gender Differences in Metabolism and Susceptibility to Environmental Exposures *Bill L. Lasley, University of California at Davis*
12:15 p.m.	Molecular Markers of Carcinogenesis: Gender Differences *Greg Cosma, Colorado State University*
12:45	Response and Discussion *Kim Boekelheide, Brown University* *Steve H. Safe, Texas A&M University* *Denise Faustman, Harvard University*

Panel III: General Discussion

1:15	Discussion of Federal Resources for Research *Committee of Federal Representatives*
1:45	Advances and Gaps in Current Research *Workshop Speakers and Committee*
2:15	Research Priorities *Workshop Speakers and Committee*
2:45	Issues and Opportunities for Agency Collaboration *Workshop Speakers and Committee*
3:15	Questions and Discussion
4:30	Adjourn

PARTICIPANTS

Donald Barnes
Staff Director
Science Advisory Board
Washington, D.C.

J. Carl Barrett
Scientific Director
National Institute of Environmental
  Health Sciences
Research Triangle Park, N.C.

Eula Bingham
Professor
Department of Environmental Health
University of Cincinnati

Kim Boekelheide
Professor
Department of Pathology
Brown University

Michael Brown
Health Scientist
Office of Women's Health
Centers for Disease Control and
  Prevention
Atlanta

Margaret Chu
Toxicologist
Environmental Protection Agency
Washington, D.C.

Gwen W. Collman
Scientific Programs Administrator
National Institute of Environmental
  Health Sciences
Research Triangle Park, N.C.

Greg Cosma
Assistant Professor
Department of Environmental Health
Colorado State University

Terri Damstra
Assistant to the Director
Center for Bioenvironmental Research
Tulane and Xavier Universities

Linda A. DePugh
Administrative Assistant
Division of Health Sciences Policy
Institute of Medicine
Washington, D.C.

Denise Faustman
Director of Immunobiology
  Laboratories
Massachusetts General Hospital
Harvard Medical School
Charlestown, Mass.

Lorraine Fitzsimmons
Program Analyst
National Institutes of Health
National Institute on Neurological
  Disorders and Stroke
Bethesda, Md.

Lynn R. Goldman
Assistant Administrator
Office of Prevention, Pesticides, and
  Toxic Substances
Environmental Protection Agency
Washington, D.C.

S. Katharine Hammond
Associate Professor
Environmental Health Sciences
School of Public Health
University of California at Berkeley

Suzanne G. Haynes
Assistant Director for Science
Office of the Secretary
Department of Health and Human
  Services
Washington, D.C.

APPENDIX D

Jean Holmes
Veterinarian
Environmental Protection Agency
Washington, D.C.

Sharon Hrynkow
Office of International Science
  Policy and Analysis
Fogarty International Center
National Institutes of Health
Bethesda, Md.

Wanda K. Jones
Associate Director for Women's
  Health
Centers for Disease Control and
  Prevention
Atlanta

Carole A. Kimmel
Senior Scientist
Food and Drug Administration
Rockville, Md.

Kenneth S. Korach
Scientific Program Director
Environmental Diseases and Medicine
  Program
National Institute of Environmental
  Health Sciences
Research Triangle Park, N.C.

Shiriki Kumanyika
Professor and Chair
Department of Human Nutrition
  and Dietetics
University of Illinois at Chicago

Bill Lasley
Professor of Reproductive Medicine
Institute for Toxicology and
  Environmental Health
University of California at Davis

C. Elaine Lawson
Research Associate
Division of Health Sciences Policy
Institute of Medicine
Washington, D.C.

Janice Longstreth
Risk Program Manager
Waste Policy Institute
Arlington, Va.

Genevieve M. Matanoski
Professor
Department of Epidemiology
Johns Hopkins University School
  of Public Health

Michael Paolisso
Director
Health and Social Analysis
International Center for Research
  on Women
Washington, D.C.

Delores Parron
Associate Director for Special
  Populations
National Institute of Mental Health
Rockville, Md.

Paul B. Phelps
Consulting Editor and Analyst
Alexandria, Va.

Vivian W. Pinn
Associate Director for Research on
  Women's Health
National Institutes of Health
Bethesda, Md.

Peter N. Riskind
Chief of Neuroimmunology
Massachusetts General Hospital
Boston

John M. Rogers
Chief
Developmental Biology Branch
Environmental Protection Agency
Research Triangle Park, N.C.

Joyce Rudick
Acting Deputy Director
Office of Research on Women's Health
National Institutes of Health
Bethesda, Md.

Stephen H. Safe
Distinguished Professor
Department of Veterinary Physiology
 and Pharmacology
Texas A&M University

Anne P. Sassaman
Director
Division of Extramural Research and
 Training
National Institute of Environmental
 Health Sciences
Research Triangle Park, N.C.

Susan Schober
Epidemiologist
National Center for Health Statistics
Hyattsville, Md.

Valerie P. Setlow
Director
Division of Health Sciences Policy
Institute of Medicine
Washington, D.C.

Ellen K. Silbergeld
Professor EPM/Director PHHE
University of Maryland Medical
 School
Baltimore

Jeanne M. Stellman
Deputy Chair
Health Policy and Management
School of Public Health
Columbia University

David H. Wegman
Professor and Chair
Department of Work Environment
University of Massachusetts

Nancy Fugate Woods
Director
Center for Women's Health Research
Professor, Family and Child Nursing
School of Nursing
University of Washington

Hal Zenick
Associate Director of Health
National Health and Environmental
 Effects Research Laboratory/EPA
Research Triangle Park, N.C.

E

Biographies of Workshop Speakers

Greg Cosma, Ph.D., is assistant professor of toxicology in the Department of Environmental Health at Colorado State University. Dr. Cosma received his B.S. degree from the University of Illinois, and his M.S./Ph.D. in pharmacology and toxicology from the University of Kansas. Following two postdoctoral fellowships at the National Cancer Institute in cellular/molecular carcinogenesis, his research has focused on the development of cancer biomarkers of exposures and susceptibility to environmental agents. He has worked closely with epidemiologists in field studies of environmentally and occupationally exposed individuals and has published numerous reports of gene-environment relationships. More recently, Dr. Cosma has explored the role of reactive oxygen species in environmental carcinogenesis and the development of biomarkers of oxidative stress. He has served on several review panels for federal cancer programs, including those of the Department of Energy, NIEHS, and the Agency for Toxic Substances and Disease Registry.

S. Katharine Hammond, Ph.D., is associate professor of environmental health sciences at the University of California, Berkeley, School of Public Health. She received her B.A. from Oberlin College, her Ph.D. in chemistry from Brandeis University, and her M.S. in environmental health sciences from the Harvard School of Public Health, where she holds an appointment as visiting lecturer in industrial hygiene. Her research has focused on assessing human exposure to complex mixtures for epidemiological studies. Among the exposures she has evaluated are those associated with work in the semiconductor industry, diesel exhaust, and environmental tobacco smoke. She served as a consultant to the EPA's Scientific Advisory Board in its review of environmental tobacco smoke; that review culminated in the publication of *Respiratory Health Effects of Pas-*

sive Smoking: Lung Cancer and Other Disorders. She is currently on the Acrylonitrile Advisory Panel for the National Cancer Institute.

Kenneth S. Korach, Ph.D., is the scientific program director of the Environmental Diseases and Medicine Program, chief of the Laboratory of Reproductive and Developmental Toxicology, and chief of the Receptor Biology Section at the NIEHS. He received his Ph.D. in endocrinology from the Medical College of Georgia in 1974. From 1974 to 1976, Dr. Korach was a postdoctoral biological chemistry research fellow at Harvard Medical School in the laboratory of the late Professor Lewis Engel. He also received a Ford research fellowship while at Harvard. In 1976 Dr. Korach joined NIEHS, where he has headed a research group investigating the basic mechanisms of estrogen hormone action in the reproductive tract and bone tissues, seeking to understand how hormonally active environmental estrogens influence physiological processes. Dr. Korach holds adjunct professorships in biochemistry at North Carolina State University and in pharmacology at the University of North Carolina Medical School. He is a recipient of the NIH outstanding performance awards, the NIH merit awards, Medical College of Georgia distinguished alumnus award, and the Edwin B. Atwood award from the Endocrine Society.

Shiriki Kumanyika, Ph.D., is professor and head of the Department of Human Nutrition and Dietetics at the University of Illinois at Chicago (UIC). She is also a professor of epidemiology in the UIC School of Public Health and the nutrition chief of service for the University of Illinois Hospital Medical Center. Dr. Kumanyika has previously held faculty positions at Cornell, Johns Hopkins, and Penn State universities. She holds a Ph.D. in human nutrition from Cornell University and master's degrees in public health (Johns Hopkins University) and social work (Columbia University) and a Bachelor of Arts in Psychology from Syracuse University. Dr. Kumanyika was a member of the IOM Committee on Legal and Ethical Issues in the Inclusion of Women in Clinical Studies. She was a member of the ORWH task forces on Opportunities for Research in Women's Health and Women in Biomedical Careers; was cochair of the Task Force on Recruitment and Retention of Women in Clinical Studies; and is an advisor to the NIH Women's Health Initiative, a very large, long-term national study of women's health. Dr. Kumanyika is actively involved in research related to nutrition epidemiology, obesity, and the health of minority populations, older populations, and women. She is the author or co-author of more than 100 publications and monographs.

Bill L. Lasley, Ph.D., is professor of environmental health and reproduction at the University of California at Davis. He received his Ph.D. from U.C. Davis and postdoctoral training at the U.C. at San Diego. Dr. Lasley was a research endocrinologist at the San Diego Zoo from 1975 until 1986, when he relocated to U.C. Davis. He has a joint appointment in the School of Veterinary Medicine, in the Department of Population Health and Reproduction and the School of

Medicine in Obstetrics and Gynecology. Dr. Lasley is currently the director of the Wildlife Health Center and associate director of the Institute of Toxicology and Environmental Health. His research work has focused on the development of noninvasive methods for investigating reproductive health, comparative reproduction, and reproductive toxicology.

Peter N. Riskind, M.D., Ph.D., is assistant professor of neurology at Harvard Medical School and assistant neurologist at Massachusetts General Hospital. Dr. Riskind is also chief of the Neuroimmunology Unit and director of the MS Treatment Program at Spaulding Rehabilitation Hospital. He received his B.A. from the University of Texas at Austin, his M.D. from the University of Texas Southwestern Medical School, and his Ph.D. in neuroendocrinology from the University of Texas Health Science Center at Dallas. His research efforts have focused on neuroendocrine regulation of prolactin secretion and on the role of hormones, including prolactin in MS. In 1977, Dr. Riskind was appointed to a National Multiple Sclerosis Society task force on Gender, MS, and Autoimmunity.

Jeanne Mager Stellman, Ph.D., is associate professor of clinical public health and deputy head of the Division of Health Policy and Management at the Columbia University School of Public Health. She received her B.S. degree from the City College of New York and her Ph.D. in physical chemistry from the City University of New York, from which she received an Alumni Distinguished Achievement Award in 1996. Her work has focused on occupational and environmental health, with a special emphasis on women's occupational health issues. Dr. Stellman was the founder and director of the Women's Occupational Health Resource Center; its papers are now housed in the Schlesinger Library at Harvard. She has published extensively in professional and lay publications and is the current editor of the journal *Women and Health*. Dr. Stellman is the editor-in-chief of the International Labor Organization's *Encyclopaedia of Occupational Health and Safety*, fourth edition, a four-volume international standard reference, to be published later this year. She was a Guggenheim Fellow and recipient of a National Cancer Institute Preventive Oncology Academic Award and has served on numerous governmental advisory panels.

F

References and Suggested Reading

Ambrosone CB, Freudenheim JL, Graham S, et al. Cigarette smoking, N-acetyltransferase 2 genetic polymorphisms, and breast cancer risk. *Journal of American Medical Association* 276(18):1494–501. 1996.

Baetjer A. *The Health Efficiency of Women at Work.* 1946

Bingham E, Morris S. Complex Mixtures and multiple agent interactions: the issues and their significance. *Fundamental and Applied Toxicology* 10:549–52. 1988.

Broersen JPJ, de Zwart, BCH, van Dijk, FJH, Meijman, TF, van Veldhoeven, M. Health complaints and working conditions experienced in relations to work and age. *Occupational and Environmental Medicine* 53:51–57. 1996.

Camann DE, Geno PW, Harding HJ, Giardino NJ, Bond AE. *Measurements to Assess Exposure of the Farmer and Family to Agricultural Pesticides.*

— Geno PW, Harding HJ, et al. A pilot study of pesticides in indoor air in relation to agricultural applications. *Proceedings of the 6th International Conference on Indoor Air Quality and Climate, Chemicals in Indoor Air, Matherial Emissions.* Indoor Air '93. Helsinki, Finland. 1993.

Centers for Disease Control and Prevention (CDC). Surveillance for anencephaly and spina bifida and the impact of prenatal diagnosis. *MMWR.* Vol. 44. Atlanta, GA: U.S. Department of Health and Human Services. 1995.

— Spontaneous abortions possibly related to ingestion of nitrate-contaminated well water. *MMWR.* Vol. 45. Atlanta, GA: U.S. Department of Health and Human Services. 1996.

— Skid-steer, loader-related fatalities in the workplace. *MMWR.* Vol. 45. Atlanta, GA: U.S. Department of Health and Human Services. 1996.

— Knowledge about folic acid and use of multivitamins containing folic acid among reproductive-aged women. *MMWR.* Vol. 45. Atlanta, GA: U.S. Department of Health and Human Services. 1996.

— Ten leading nationally notifiable infectious diseases. *MMWR.* Vol. 45. Atlanta, GA: U.S. Department of Health and Human Services. 1996.
— surveillance summaries. *MMWR.* Vol. 45. Atlanta, GA: U.S. Department of Health and Human Services. 1996.
— Asthma mortality and hospitalization among children and young adults. *MMWR.* Vol. 45. Atlanta, GA: U.S. Department of Health and Human Services. 1996.
Devesa SS, Blot WJ, Stone BJ, Miller BA, Tarone RE, Fraumeni JF. Recent cancer trends in the United States. *Journal of the National Cancer Institute.* 87(3):175–182. 1995.
EPA. "Strategic plan for the office of research and development." Washington, D.C. 1995.
— CDC, and Agency for Toxic Substances and Disease Registry. *Inventory of exposure-related data systems sponsored by federal agencies.* EPA/600/R-920/078. 1992
— *Nonoccupational Pesticide Exposure Study (NOPES),* EPA/600/3-90/003. EPA, Washington, D.C. 1990.
Frankenhaeuser M. The physiology of sex differences as related to occupational status. In *Women, Work and Health*: Frankenhaeuser M, Lundberg U, and Chesney M, editors. Plenum Press. New York. 1989.
Geno PW, Camann DE, Harding HJ, Villalobos K, Lewis RG. Handwipe sampling and analysis procedure for the measurement of dermal contact with pesticides. *Archives of Environmental Contamination Toxicology* 30:132–38. 1996.
Grassman JA. Obtaining information about susceptibility from the epidemiological literature. *Toxicology* 111(1–3):253–270. 1996.
Hattis D. Human interindividual variability in susceptibility to toxic effects: from annoying detail to a central determinant of risk. *Toxicology* 111(1–3):5–14. 1996.
Hsu JP, Camann DE, Schattenberg H III, et al. *New Dermal Exposure Sampling Technique.* Raleigh, NC: EPA. 1990.
IOM. 1997. Liverman C, Ingalls C, Fulco C, and Kipen H. *Toxicology and Environmental Health Information Resources.* Washington, D.C.: National Academy Press. 1997.
— 1996. Berns K, Bond E, Manning R, editors. *Resource Sharing in Biomedical Research.* Washington, D.C.: National Academy Press. 1996.
— 1994a. Andrews L, Fullerton J, Holtzman N and Motulsky A, editors. *Assessing Genetic Risks.* Washington, D.C.: National Academy Press. 1994.
— 1994b. Lynch B and Bonnie R, editors. *Growing Up Tobacco Free: Preventing Nicotine Addiction in Children and Youths.* Washington, D.C.: National Academy Press. 1994.
Khoury MJ, Flanders WD, Greenland S, Adams MJ. On the measurement of susceptibility in epidemiologic studies. *American Journal of Epidemiology* 129(1):183–190. 1989.
Kurtzke JF. Geography in multiple sclerosis. *Journal of Neurology* 215(1):1–26. 1977.
— Beebe GW, Norman JE, Jr. Epidemiology of multiple sclerosis in U.S. veterans: 1. race, sex, and geographic distribution. *Neurology* 29(9):1228–1235. 1979.
— Page WF. Epidemiology of multiple sclerosis in US veterans: VII. Risk factors for MS. *Neurology* 48(1):204–213. 1997.
Lewis RG, Fortmann RC, Camann DE. Evaluation of methods for monitoring the potential exposure of small children to pesticides in the residential environment. *Archives of Environmental Contamination and Toxicology* 26:37–46. 1994.

— Roberts JW, Chuang JC, Camann DE, Ruby MG. Measuring and reducing exposure to the pollutants in house dust (Letter to the Editor). *American Journal of Public Health* 85(8):1168. 1995

Liverman C, Ingalls C, Fulco C, Kipen H, editors. *Toxicology and Environmental Health Information Resources*. Washington, D.C.: National Academy Press. 1997.

Lukas SE, Sholar M, Lundahl LH, et al. Sex differences in plasma cocaine levels and subjective effects after acute cocaine administration in human volunteers. *Psychopharmacology* 125(4):346–354. 1996.

Mitchell FL, editor. *Multiple Chemical Sensitivity: A Scientific Overview*. Princeton, NJ: Princeton Scientific Publishing Co., Inc. 1995.

National Academies of Science and Engineering and Institute of Medicine. *Allocating Federal Funds for Science and Technology*. Washington, D.C.: National Academy Press. 1995.

NCHS. *Health, United States, 1995 Chartbook*. Hyattsville, MD: Public Health Service. 1996.

NIOSH. *Reproductive Hazards in the Workplace Bibliography*. Cincinnati, OH: U.S. Department of Health and Human Services. 1994.

NIEHS. Announcement on internet: http//www.niehs.nih.gov/dirosd/policy/egp. 1997.

National Research Council (NRC). *Monitoring Human Tissues for Toxic Substances*. Board on Environmental Studies and Toxicology, Commission on Life Sciences. Washington, D.C.: National Academy Press. 1991.

— *Linking Science and Technology to Society's Environmental Goals*. Washington, D.C.: National Academy Press. 1996.

Ness RB, Kuller LH. Women's health as a paradigm for understanding factors that mediate disease. *Journal of Women's Health*, 6(3):329–336. 1997.

Nigg HN, Beier RC, Carter O, et al. Exposure to pesticides. Wilkinson SR, Baker CF, editors. *The Effects of Pesticides on Human Health*. Vol. XVIII. Springfield, VA: Risk Focus Versar, Inc., 35–130. 1990.

Nishioka MG, Burkholder HM, Brinkman MC, Gordon SM, Lewis RG. Measuring transport of lawn-applied herbicide acids from turf to home: correlation of dislodgeable 2,4-D turf residues with carpet dust and carpet surface residues. *Environmental Science and Technology* 30(11):3313–20. 1996.

ORWH. "Scientific research program." U.S. Department of Health and Human Services. 1996

Perera FP. Molecular epidemiology: insights into cancer susceptibility, risk assessment, and prevention. *Journal of National Cancer Institute* 88(8):496–509.

Petersdorf R, Page W, Thaul S, editors. *Interactions of Drugs, Biologicals and Chemicals in the U.S. Military Forces*. Washington, D.C.: National Academy Press. 1996.

Peterson and Wiley, L 1995.

Preston RJ. Interindividual variations in susceptibility and sensitivity: linking risk assessment and risk management. *Toxicology* 111(1–3):331–341. 1996.

Roush, W. Chimp Retirement Plan Proposed. *Science* Vol. 227:471. 1997.

Russo J and Russo, IH. Toward a unified concept of mammary carcinogenesis. In *Etiology of Breast and Gynecological Cancers*. Anonymous, pp. 1–16, Wiley-Liss, Ind. 1997.

Safe SH. Xenoestrogens and breast cancer. *New England Journal of Medicine*. Vol. 337:1303–1304. 1997.

— Environmental and dietary estrogens and human health: is there a problem? *Environmental Health Perspectives* 103:346–351. 1995.

United States Social Security Administration (SSA). *Disability*. 1995. Pamphlet.

— *Disability Evaluation Under Social Security*, SSA Publication No. 64-039 U.S. Department of Health and Human Services. 1994.
— *What You Need to Know When You Get Disability Benefits*. Social Security Administration. 1996. Pamphlet.
Wallace LA. A decade of studies of human exposure: what have we learned? *Risk Analysis* 13(2):135–43. 1993.
— Environmental exposure to benzene: an update. *Environmental Health Perspectives* 104: Supplement 6:1129–36. 1996.
— Indoor particles: a review. *Journal of The Air and Waste Management Association* 46:98–126. 1996.
— The California TEAM study: breath concentrations and personal exposures to 26 volatile compounds in air and drinking water of 188 residents of Los Angeles, Antioch, and Pittsburg, CA. *Atmospheric Environment* 22(10):2141–63. 1998.
— Cancer Risks from Organic Chemicals in the Home. *Proceedings from an APCA International Specialty Conference on Environmental Risk Management*. 1988.
— Exhaled breath as an indicator of recent exposure to volatile organic compounds. Presentation at the 80th Annual Meeting of APCA. 1987.
— The Total Exposure Assessment Methodology (TEAM) Study. *EPA Project Summary*. EPA/600/S6-87/002 (1987).
— The Total Exposure Assessment Methodology (TEAM) Study: Summary and Analysis, Vol. I. Washington, D.C.: EPA, Office of Acid Deposition, Environmental Monitoring and Quality Assurance. 1987.
— Pellizzari ED, Gordon SM. Organic chemicals in indoor air: a review of human exposure studies and indoor air quality studies. *Indoor Air and Human Health*. Kaye RB, Gammage SV, 361–78. Lewis Publishers, Inc.
— Ziegenfus RC. Comparison of carboxyhemoglobin concentrations in adult nonsmokers with ambient carbon monoxide levels. *Journal of the Air Pollution Control Association* 35(9):944–49. 1985.
— Pellizzari ED. Personal air exposures and breath concentrations of benzene and other volatile hydrocarbons for smokers and nonsmokers. *Toxicology Letters* 35:113–16. 1986.
— The exposure of the general population to benzene. *Cell Biology and Toxicology*. 5(3):297–314. 1989.
— Pellizzari ED, Hartwell TD, Davis V, Michael LC, Whitmore RW. The influence of personal activities on exposure to volatile organic compounds. *Environmental Research* 50:37–55. 1989.
— Pelizari E, Wendel C. Total volatile organic concentrations in 2700 personal, indoor, and outdoor air samples collected in the U.S. EPA TEAM Studies. *Indoor Air* 4:465–77. 1991.
— Buckley E, Pellizzari E, Gordon S. Breath measurements as volatile organic compound biomarkers. *Environmental Health Perspectives*. 104: Supplement 5:861–69. 1996.
van Dormolen M, Hertog CA, can Dijk FJ, Kompier MA, Fortuin R. *The Quest for Interaction: Studies on Combined Exposure*. International Archives of Occupational Environmental Health. 62:279–287. 1990.
Welch LS. *Case Studies in Environmental Medicine: Reproductive and Developmental Hazards*. U.S. Department of Health and Human Services. 1993.
Whitmorem RW, Immerman FW, Camann DE, Bond AE, Lewis RG, Schaum JL. Non-Occupational exposures to pesticides for residents of two U.S. cities. *Archives of Environmental Contamination and Toxicology* 26:47–59. 1994.

G

Committee Biographies

Nancy Fugate Woods, Ph.D. (Chair), is associate dean for research and professor of family and child nursing at University of Washington School of Nursing. She received a B.S. in nursing from the University of Wisconsin, Eau Claire, in 1968; an M.N. from the University of Washington in 1969; and a Ph.D. in epidemiology from the University of North Carolina, Chapel Hill, in 1978. Since the mid-seventies, Dr. Woods has provided leadership in the development of women's health as a field of study in nursing science. Her early research focused on the relationship of women's social environments and health. Since the late 1970s Dr. Woods has led several large research projects focusing on women's experiences with perimenstrual symptoms. With collaborators at Duke University and the University of Washington, she conducted the first prevalence study of perimenstrual symptoms among U.S. women. Subsequent research focused on women's social environments, stress response, and ovarian hormones in the etiology of menstrual cycle symptoms. In collaboration with colleagues at the University of Washington, Dr. Woods established in 1989 the Center for Women's Health Research, focusing on women's health across the lifespan. Current research focuses on women in midlife health, and health-seeking behavior patterns. She is currently involved in projects focusing on menopause, including women's decisions about using hormone replacement therapy. Dr. Woods is active in professional organizations, having served as president of the American Academy of Nursing and the Society for Menstrual Cycle Research. She is also a member of the National Advisory Council on Nursing Research for the National Institute on Nursing Research and the Women's Health Task Force at NIH. She received the American Nurses Foundation Distinguished Contribution to Nursing Research Award and is a member of the Institute of Medicine, National Academy of Sciences.

Eula Bingham, Ph.D., is a professor in the Department of Environmental Health at the University of Cincinnati College of Medicine. She was vice president for research at the University of Cincinnati for nine years. Dr. Bingham served as assistant secretary of labor for occupational safety and health in the Carter Administration. She has done research in chemical carcinogenesis and toxicology and is currently involved in research on surveillance methodologies for construction workers in the nuclear industry. Dr. Bingham has been a leader in policy development to protect women from discrimination based on hazardous workplace exposures. She currently serves as chair of the Board of Scientific Counselors for the Agency for Toxic Substances and Disease Registry and chair of the Department of VA's Persian Gulf Advisory Committee. Dr. Bingham is also a member of the National Toxicology Program Board and the IOM.

Kim Boekelheide, Ph.D., is professor of pathology and laboratory medicine at the Brown University School of Medicine. He received his B.A. from Harvard University and M.D. and Ph.D. from Duke University. His research has examined fundamental molecular mechanisms by which toxicants induce testicular injury. Current projects include an assessment of cytoskeletal perturbation in testicular injury and the biological basis of irreversible testicular atrophy, focusing on pathways of germ cell apoptosis. He has been continuously funded by the NIEHS since 1985 and has received several awards, including a Burroughs Wellcome Toxicology Scholar Award (1994–1999). He has served as a member (1990–1995) and Chair (1993–1995) of the Toxicology Study Section of the Division of Research Grants, NIH.

Denise Faustman, M.D., Ph.D., is associate professor of medicine at Harvard Medical School and director of the immunobiology laboratory at Massachusetts General Hospital (MGH). She earned her M.D. and Ph.D. at Washington University School of Medicine. She did her internal medicine and endocrinology training at MGH, where she currently directs the immunobiology laboratory. Her research focuses on transplantation, autoimmunity, and the disparity in the incidence of autoimmune diseases between men and women. Dr. Faustman is the author of many articles in *Science* and PNAS and in 1991 presented data implicating the antigen-presenting cells of peripheral blood as the "bad" eductor cells in most autoimmune diseases. This year, she identified a new mutation that may be central in the gender-controlled expression of disease.

Stephen H. Safe, D.Phil., is a distinguished professor in the Department of Veterinary Physiology and Pharmacology at Texas A&M University. He received his B.Sc. and M.Sc. at Queen's University in Canada and a D.Phil. from Oxford University in England. His research is focused on several areas, including biochemical, toxic, and genotoxic responses of halogenated and polynuclear aromatic hydrocarbons; the adverse effects of endocrine-disrupting compounds; the regulation of estrogen-induced gene expression and crosstalk with AH receptors; and the development of AH receptor-based drugs for breast cancer treatment. He is a member of several committees and currently serves as a councilor for the Society of Toxicology.

David H. Wegman, M.D., is professor and chair, Department of Work Environment at the University of Massachusetts at Lowell. He received his B.A. from Swarthmore College and his M.D. and M.Sc. from Harvard University. Dr. Wegman has focused his research on epidemiological studies of occupational respiratory disease, musculoskeletal disorders, and cancer and has published over 100 articles in the scientific literature. He has also written on public health and health policy issues, such as hazard and health surveillance, methods of exposure assessment for epidemiologic studies, the development of alternatives to regulation, and the use of participatory methods to study occupational health risks. He is coeditor with Dr. Barry Levy of one of the standard textbooks in the field of occupational health, *Occupational Health: Recognition and Prevention of Work-Related Disease*, the third edition of which was published in 1995. His recent work has focused on developing methods to study subjective outcomes, such as respiratory or irritant symptoms reports, and on the health and safety risks among construction workers involved in the building of the Third Harbor Tunnel and the underground Central Artery in Boston.

STAFF

Valerie Petit Setlow, Ph.D., is the director of the Division of Health Sciences Policy, IOM. In this capacity she is responsible for the development of public policy activities related to biomedical research, including fundamental science and clinical research; infrastructure to support research; drug development and regulation; education, training, and mentoring of health professionals; and the ethical, legal, and social implications of biomedical advances. Dr. Setlow received her B.S. in chemistry from Xavier University in 1970 and her Ph.D. in molecular biology from The Johns Hopkins University in 1976. Dr. Setlow has conducted research in molecular hematology and virology and has had a distinguished career in government, serving in many seven different staff positions, including director of the Cystic Fibrosis Research Program at NIH, senior policy analyst in the Office of the Assistant Secretary for Health, and as acting director of the National AIDS Program Office. She also holds an adjunct appointment at Howard University in the Department of Community Health and Family Medicine, where she teaches medical bioethics.

C. Elaine Lawson is a research associate of the Division of Health Sciences Policy, IOM. In this capacity, she is responsible for providing background papers, staff analyses, and reference research for study directors and/or the division director. Most of her work has centered around ethical, legal, and social implications of biomedical advances and K–12 health and science education. Ms. Lawson received her B.S. in health and physical education from James

Madison University in 1978 and her M.S. in exercise science and health from George Mason University in 1994. Ms. Lawson has conducted research in health education policy and public genetics education. She began her IOM career in 1989 as a senior project assistant. She became a research assistant in 1992 and a research associate in 1994.

Linda DePugh is the administrative assistant for the Division of Health Sciences Policy, IOM. Ms. DePugh has been a member of the National Academy of Sciences staff for over 25 years. She served as administrative assistant for the Division of Health Promotion and Disease Prevention prior to joining the Division of Health Sciences Policy in 1994. Ms. DePugh provides administrative assistance to the Board on Health Sciences Policy and the division by coordinating specific tasks which are crucial to the progress and completion of program activities. She obtained her associate's degree from Durham Business College.

Jamaine Tinker is a Financial Associate with the IOM. She provides support throughout the life research projects by completing financial and administrative responsibilities. She works closely with program staff to prepare proposal and working budgets, cost projections, and financial reports and analyses. Also, in this capacity, she often serves as the liaison between the division and other academy offices, such as Contracts and Grants, Accounting, Purchasing, Payroll, Travel Services, and Human Resources. Ms. Tinker earned a business administration certificate from Georgetown University in 1994 and a B.A. from Wittenberg University in 1987, where she majored in Spanish and minored in math.